Terrestrial Environmental Sciences

Series Editors

Olaf Kolditz, Helmholtz Centre for Environmental Resea, Leipzig, Germany

Hua Shao, and Natural Resources, Federal Institute for Geosciences, Hannover, Germany

Wenqing Wang, UFZ, Environmental Informatics, Helmholtz Centre for Environmental Resea, Leipzig, Germany

Uwe-Jens Görke, Environmental Informatics, Helmholtz Centre for Environmental Resea, Leipzig, Germany

Sebastian Bauer, University of Kiel, Institute of Geosciences, Kiel, Germany

Understanding the function and evolution of terrestrial environmental systems is fundamental to many environmental aspects investigated in the geo- and hydrosciences. The terrestrial environmental systems under investigation here range from the geosphere and its related water cycle to associated matter fluxes and biogeochemical transformations. Modelling is important for system characterization and understanding as well as describing potential paths of terrestrial environmental systems. Benchmarking builds a bridge to experimental studies and provides a methodology for model validation. Moreover, benchmarking and code comparison foster community efforts. This book series invites contributions in fundamental and applied aspects in terrestrial environmental sciences as well as in other related fields to promote interdisciplinary approaches.

More information about this series at http://www.springer.com/series/13468

Max Dohmann · Martin Grambow ·
Yonghui Song · Paul Wermter

Editors

Chinese Water Systems

Volume 4: Applied Water Management
in China

 Springer

Editors
Max Dohmann
Research Institute for Water and Waste
Management at RWTH AC
Aachen, Germany

Martin Grambow
Water Management and Soil Protection
Bayerisches Staatsministerium für Umwel
Prien am Chiemsee, Bayern, Germany

Yonghui Song
Chinese Research Academy of
Environmental Sciences (CRAES)
Anwai Beiyuan, Beijing, China

Paul Wermter
Ministry of Environment, Energy, Food
and Forestry of Rhineland-Palatinate
Mainz, Germany

ISSN 2363-6181   ISSN 2363-619X   (electronic)
Terrestrial Environmental Sciences
ISBN 978-3-030-80236-3   ISBN 978-3-030-80234-9   (eBook)
https://doi.org/10.1007/978-3-030-80234-9

Introduction

For sustainable water management, China is facing triple challenges of water resources, water environment, and water ecology. China's per capita water resources are only 2,300 cubic metres, which is only 1/4 of the world average. Water resources are unevenly distributed in time and space, with more in the south and less in the north, more in summer and less in winter, which has become a restrictive factor for economic and social development. China is also facing the challenge of water pollution. The seven major rivers, Songhuajiang, Liaohe, Haihe, Yellow River, Huaihe, Yangtze River, and Pearl River, and the three regions of the south-east, south-west, and north-west, all have different levels of water pollution problems; especially, the Haihe, Liaohe, Yellow River, and Huaihe are more polluted; lake eutrophication is more common, especially in Lake Taihu, Lake Chaohu, and Lake Dianchi. Water pollution has exacerbated water shortages due to water quality. The rapid economic and social development and the over-exploitation of river and lake basins have also caused damage to the water ecology, increasing the difficulty of non-point source pollution and water environment management.

In response to the water management challenges, China has continued to reform and innovate, and put forward a strategy of "prioritizing water conservation, spatial balance, systematic governance, and two-handed efforts" to manage water. In 2015, the revised "Environmental Protection Law" was implemented. Increased penalties and rigid constraints for environmental violations, and at the same time initiated the implementation of the "Water Pollution Prevention and Control Action Plan", comprehensively considering water saving, water conservation, economic and social development, management and sustainable use of water resources, with the goal of comprehensively improving the safety level of water resources, water environment and water ecology; in 2018, the revised "Water Pollution Prevention and Control Law" was started to implement, and innovative systems such as the River Chief System and the Lake Chief System have been established nationwide.

Scientific water management strategies require solid scientific and technological support. The Major Science and Technology Program for "Water Pollution Control and Governance" (hereinafter referred to as the "Major Water Programme") covering three Five-Year Plans from 2007 to 2020 aims to solve the bottleneck technological problems of water pollution control and governance of China. The technologies for industrial water pollution control, urban water pollution control, agricultural and rural non-point source pollution control, drinking water safety guarantee, basin water environmental management, water ecological restoration, etc., have been developed, and technological engineering demonstrations in the Taihu, Chaohu, Dianchi, Haihe, Liaohe, Huaihe basins, etc., have been carried out. It is planned to build up two technological systems of China for water pollution control and water environment management at basin level, so as to improve the modernization level of China's water governance and management.

During the implementation of the Major Water Programme, in addition to independent innovation, attention was also paid to the international cooperation, especially technical exchanges with Germany. In 2011, China and Germany signed a joint statement on the "Clean Water Research and Innovation Programme"; on 7 May 2015, the Sino-German Water Cooperation Launch Meeting was held. The Ministry of Science and Technology (MoST) of China and the German Federal Ministry of Education and Research (BMBF) signed a "Major Water Programme" cooperation declaration and officially launched three German–Chinese water research cooperation projects: SIGN, "Drinking water quality assurance in the Lake Taihu basin"; SINOWATER, "Good water governance, water resources management and innovative technologies (Liaohe and Dianchi)"; and UrbanCatchments, "Water resources management in urban watersheds". These three projects initiated by the German side have formed a good connection with the 12th Five-Year Plan projects of the Chinese Major Water Programme. The German–Chinese cooperation has provided a new approach and reference to solve the water pollution problem in the main basins of China, and both sides have benefited from the cooperation in terms of technology innovation, technology marketization, and solutions for water challenges.

The SINOWATER project takes into account the large-scale heavy chemical industry polluted river, "Liaohe", and the plateau eutrophic lake, "Dianchi", focusing on the core city of the north-east old industrial base, "Shenyang", and the metropolis of the Dianchi basin, "Kunming". In both basins and urban areas, comprehensive research including good water governance technology, wastewater treatment plant upgrading technology, pharmaceutical wastewater treatment technology, sustainable sludge treatment technology, etc., was carried out. These research contents are exactly the urgent needs in the above basins and areas. In the organization of the research team, for the Liaohe basin and Shenyang area, the Research Institute for Water and Waste Management at RWTH Aachen (FiW) e.V., Bavarian State Ministry of the Environment and Consumer Protection, Technical University of Munich, RWTH Aachen University and the related German industrial partners, the Chinese Research Academy of Environmental Sciences, the Shenyang Research Academy of Environmental Sciences, and the Guodian Northeast Environmental Industry Group Co., Ltd. formed a German–Chinese cooperation team with close integration of "Industry, university, and research institution", showing the advantages of collaborative innovation. Practice has shown that this organizational method is essential for the output of innovative results, the

construction of systematic water governance plans, and the application and transformation of technologies. After several years of collaborative research, the projects
of both China and Germany have been successfully completed. The systematic
summary of SINOWATER's research results by publishing a book would be of
great significance for the implementation of the ongoing second phase of the
German–Chinese cooperation project and the final accomplishment of the Chinese
Major Water Programme. As the coordinator of the Chinese side of the
SINOWATER project, I sincerely thank the researchers from both China and
Germany, especially Professor Dr.-Ing. Max Dohmann, Professor Dr.-Ing. Martin
Grambow, Dr.-Ing. Paul Wermter, and many other colleagues for their hard work
and sincere cooperation; I also sincerely thank the leaders and professional man-
agers of the related governmental agencies from both China and Germany, for their
strong support and enthusiastic help.

Yonghui Song

German–Chinese cooperation in the field of environmental research has a long
tradition and began over 30 years ago. This was based on agreements between the
German Federal Ministry of Education and Research (BMBF) and the Chinese
Ministry of Science and Technology (MoST). The main motives for this cooperation were the adapted application of German environmental technologies in China
and the associated expectation of German economic benefits. From the very
beginning, the joint projects in water research were given special priority.

The German research programme "International Partnerships for Sustainable
Climate Protection and Environmental Technologies and Services (CLIENT)"
launched in 2010 and the signing of the Chinese–German "Clean Water Research
and Innovation Programme" by the research ministers of both countries in 2011 set
the course for intensified cooperation in bilateral water research.

In addition, the CLIENT II programme of the German Federal Ministry of
Education and Research (BMBF) was launched in 2015, which for the first time
also required coordination with the Chinese Ministries of Environment (MEP) and
Housing and Urban and Rural Development (MoHURD) as part of the Chinese
"Major Program of Science and Technology for Water Pollution Control and
Governance", which has been running since the beginning of the 11th Chinese
Five-Year plan in 2006. This was important because the two ministries mentioned
above are responsible for implementing the results of the research projects in
Chinese practice.

The Chinese Major Water Programme focused on 10 major and particularly
polluted Chinese water resources, namely five major river basins and five major
lakes. The involvement of German research institutions and industrial partners in
improving the water management conditions for one of these river basins and for
three of the lakes resulted from three collaborative research projects started in 2015.
In line with the orientation of the Chinese Major Water Programme, these projects
included technological aspects as well as management and good governance
approaches. These three German–Chinese cooperation projects started with the
inauguration event held on 07.05.2015 in Beijing where a joint declaration between

the Chinese Ministry of Science and Technology (MoST) and German Federal Ministry of Education and Research (BMBF) was signed. On the Chinese side, the Chinese Research Academy of Environmental Sciences in the person of its Vice-President Prof. Yunghui Song played a coordinating role. Compared with earlier German–Chinese research projects, the three joint projects had the advantage that on the Chinese side, research groups had already proven themselves in the Major Water Programme were established.

This volume reports on the joint project "SINOWATER Good Water Governance, Management and innovative Technologies to improve Water Quality in two important Chinese Waters", one of the three joint projects mentioned above. The SINOWATER joint project covered the Liao River and Dianchi Lake with the metropolitan areas of Shenyang and Kunming. Using sustainable and innovative German technologies and successful experience with applied conceptual approaches, SINOWATER aimed to promote the river and lake resources of the Liao and Dianchi areas and to develop and optimize improved water management.

One objective of SINOWATER was the development of modified structures and organizational measures to improve analytical and decision-making capabilities in the normative and operational management of the water sector on the basis of cooperative, participatory, and specifically ecological research approaches. A further objective was to reduce the pollution of the water bodies under consideration by wastewater discharges. In addition, the aim was to contribute to the continuation of a long-term master plan for the Dianchi area.

The joint project SINOWATER was supplemented in 2016 by two project parts in the cities of Suzhou and Jiaxing. These were technological approaches to ensure very low nutrient concentrations in the effluents of wastewater treatment plants with a high industrial wastewater content and investigations with a new German equipment for cleaning and inspection of sewers.

In 2018, the joint project SINOWATER was completed. This volume summarizes the main results of the joint research project.

Max Dohmann

The Bavarian Water Administration supports since more than twenty years international projects. Why? There is no place, no way to learn fast and holistic about water than in the dialog with experts from other countries and other climate and cultural zones.

This approach is important for all stakeholders. The meaning of water for any part of our life and out environment is widely underestimated: water seems to be the most political resource. Water is the unique common, which has to be treated wisely in the sense of present and future generations. Mismanagement of water resources is jeopardizing the stability of any community, of any state. If there is a basic need of good water management, then there is also a basic need in a good water administration.

In this respect, it must be the basic concern of every water administration to learn together and from each other in order to fulfil this task in the best possible way. The German–Chinese joint research project SINOWATER, whose results and findings are published for the first time as a whole in this book, has taken this path.

Actually, we are in a situation of fundamental growth of knowledge about our basic living conditions. We live in the age of the Anthropocene. Man is the determining factor of this geological epoch. Due to our rapid technical and scientific progress, we have intervened in the earth system and the water balance (e.g. dams and hydroelectric power plants). This has partly brought us a great increase in prosperity. On the other hand, there are increasing signs that unconsciously induced side effects of our actions are endangering our prosperity. The decline in biodiversity, climate change, or ubiquitous chemical substances are unmistakable signs of this. In addition, there are other challenges such as massive demographic changes.

Water policy and water management will therefore be no less important in future. They are permanent tasks that must be constantly developed further.

This also requires the further development of suitable indicators, not least in order to make highly complex interdependences in the water sector accessible for status description, control, and monitoring. Participation and communication are also ultimately based on this.

Within the framework of the research project, we have therefore also intensively dealt with the topic of indicators and have taken up the challenge of developing not only purely technical indicators, but also non-technical ones, such as for the description of the quality of life or for administrative structures itself. Ultimately, this has resulted in a separate chapter in this book.

With this book, the SINOWATER research project is now finally concluded. We hope that through this book we will succeed in generating the real added value of our research work, namely the integration of the results into practical water management and the inspiration for future research activities.

Martin Grambow

Contents

About the Editors

Yonghui Song is Professor and Vice-President of the Chinese Research Academy of Environmental Sciences (CRAES), which is affiliated to the Ministry of Environmental Protection (MEP) of China. He obtained his Ph.D. degree in environmental science from the Research Center for Eco-Environmental Sciences of the Chinese Academy of Sciences in 1999 and his Dr.-Ing. in environmental engineering from the University of Karlsruhe (TH), Germany, in 2003. His research focuses on water pollution control technologies, and regional- and basin-level water environment management. He has undertaken over 20 national or ministerial/provincial scientific research projects as the principal investigator, published over 300 journal papers, and obtained more than 20 patents. He received the "Outstanding Research Team Award" of the "11th Five-Year Plan" of the Ministry of Science and Technology (MoST) of China in 2012 and the "Award of MEP for Science and Technology" in 2016. He was selected as the Youth Innovation of Science and Technology Leading Talent by MoST of China in 2012 and was selected as the Innovation of Science and Technology Leading Talent by Organization Department of the CPC Central Committee General Office of China in 2013.

Max Dohmann is Professor Emeritus of the RWTH Aachen University and Member of the Board of Research Institute of Water and Waste Management at RWTH Aachen University. His research interests concern general environmental aspects and especially urban drainage, wastewater treatment, water quality management, waste, and waste management. After studying civil engineering at the RWTH Aachen University and several years of engineering work in a water management association, he received his doctorate from the University of Hanover in 1974. He then continued his responsible practical work at the water management association and became Professor at the University of Essen in 1983. From 1987 to 2004, he worked as Professor and Director of the largest German institute for urban water management at the RWTH Aachen University. For several years, he was Member of the German Federal Government's Council of Experts on Environmental Issues and Long-standing Member of the Governing Board of the International Water Association (IWA). For many years, he is Visiting Professor at Tsinghua University in Beijing and at Sichuan University in Chengdu. In 2015, the German Ministry of Research appointed him as German spokesperson for the first three German collaborative research projects in the Chinese Major Water Programme.

Martin Grambow is Director General of "Water Management and Soil Protection" at the Bavarian Ministry of Environment and Consumer Protection. His responsibilities comprise all aspects of stewardship for the commons "water" and "soil", including regulation, monitoring, and management. In this respect, he also represents the state Bavaria in national and international bodies. Among others, he chairs the German River Basin Management Community Danube and is delegate in the International Lake Constance Conference and the International Commission for the Protection of the Danube River. In 2020 and 2021, he is Chairman of the

Federal/States Work Group on Water Issues (LAWA) and the Work Group on Soil Protection (LABO).

In the late 90th, he founded the stat unit "Technology Transfer Water (TTW)". Via TTW, the Bavarian State water administration supports the exchange of knowledge in water administration and in water technologies. He contributes to international activities carried out by various institutions including World Bank, OECD, European Union, or Deutsche Gesellschaft für internationale Zusammenarbeit (GIZ). The Asia Region is a main focus of his international partnerships, dialogues, and exchanges.

He is Member of several institutions and foundations, notably the European Academy of Science and Arts and their "International Expert Group on Earth System Preservation" (IESP), and the "Bavarian Water Foundation". He is Member of the Managing Board of the German Association for Water, Wastewater and Waste (DWA), of the Bavarian Environment Cluster (Umweltcluster Bayern), and specially notably since 2018 Member of the International Scientific Advisory Committee (ISAC) for the Chinese Research Academy of Environmental Sciences (CRAES).

He studied civil engineering at TUM Technische Universität München. He graduated in 1986 (Diplom-Ingenieur), achieved his doctor's degree in 2005, was appointed Honorary Professor in 2012, and teaches International Waterpolicies and Waterrights at TUM.

He is Author of various books and articles primarily focusing on the application of concepts for sustainable (water) resource management. He also has co-edited the books "Sustainable Risk Management" and "Global Stability through Decentralization?" in the Springer series "Strategies fur Sustainability".

Paul Wermter received his diploma as Environmental Engineer at University of Rostock 1999. Shortly after, he joined Planungsbüro Koenzen, one of Germany's leading consultants in the field of river rehabilitation. Ever since, he was concerned with hydraulics and especially with the hydrology of rivers in regard to river typology. R&D took him to water quality as well as point and non-point source pollution modelling and control. Early 2006, he committed himself fully to applied R&D at Research Institute of Water and Waste Management at RWTH Aachen (FiW). His research priorities comprised economic aspects of European Water Framework Directive, financing/pricing of water management, micro-pollutants within aquatic environment, and urban and climate impacts on the hydrology and the flow regime of rivers. In 2009, he was appointed Scientific Head of Integrated Water Resources Management at FiW and in 2014 Head of International Cooperation China. He was Coordinator of national and international joint R&D projects funded by BMBF like SINOWATER. Today, he is working at the Ministry of Environment of Rhineland-Palatinate, responsible for digitization in the water sector of that federal state of Germany.

Chapter 1
Improving Water Management and the Water Management System in the Liao River Basin and Lake Dian

Ke Chen and Markus Disse

1.1 Structural Indicators

With increased political and academic concerns, the term "water governance" has been debated that it responds to challenges of sustainable development [42]. The Organization for Economic Co-operation and Development (OECD) defines water governance as a "range of political, institutional and administrative rules, practices and processes (formal and informal) through which decisions are taken and implemented, stakeholders can articulate their interests and have their concerns considered, and decision makers are held accountable for water management" [35]. Good water governance thus depends on several significant factors, including strong legal, policy and regulatory framework, effective administrative and institutional framework that facilitates collaboration among various water organizations; civic determinations for sustaining good water environment and adequate investments [10]. Actually, the term of Integrated Water Resources Management (IWRM) can be described by technical and management approaches, wherein the management approaches can be divided into "Good Water Governance Structures, Water Rights, Financing, Communication, Human Factors and Cultural Impacts" [11]. Having a closer look to the Good Water Governance Structure, it comes out that there in absolutely no common design of water institutions but in the opposite any culture has developed very individual ways to organize the water issue. This is remarkable, because the water issue itself globally is not that different: water is needed by any life, is a typically regional or temporally changing resource, from drought seasons to abundant up to floods and water is running downhill, which connects needs of any living thing in the river basin in a competitive way. So, the water governance is a result of an institutional, legal and cultural organization of any existing state or society, which is at least

K. Chen · M. Disse (✉)
Chair of Hydrology and River Basin Management, TUM Department of Civil, Geo and Environmental Engineering, Technical University of Munich, Munich, Germany
e-mail: markus.disse@tum.de

M. Dohmann et al. (eds.), *Chinese Water Systems*, Terrestrial Environmental Sciences,
https://doi.org/10.1007/978-3-030-80234-9_1

1

complicated if not complex because it involves actually any form of state structure, from the highest level (State) down to the lowest subsidiary level (communities, clans, families) [12]. This factor water institutional framework (water governance structure) plays a strong role in affecting the overall water governance performances. While effective water institutional framework helps to shape the choices made within water resource management, it reflects the stakeholders with their respective goals, features the fractures and balances of powers, determines the controlling and collaborative mechanism among water relevant organizations and ensures collective actions against existing and new challenges [6, 38]. Thus, arguments have been made that water resource governance depends strongly on placement of institutional arrangements, and water institutional framework is of vital importance in ensuring water security [14, 41]. Therefore, understanding the relevant organizations and their institutional dynamics could facilitate the stakeholders and decision makers to conduct effective water management measures.

1.1.1 Development of Innovative 3D Water Governance Models for Kunming City and Bavaria

1.1.1.1 Introduction

It has been recognized that any institutional framework, including water sector, could reflect its several intrinsic dynamic aspects, such as the formal and informal scheme of relationships, patterns and channels of communication and operating processes for routine and exceptive management tasks [1, 19]. If taking the structure of the organization of the society as the "hardware", it also requires understanding the "software", meaning the written and unwritten rules of cooperating and teamworking to understand the management. Thus, these management processes of various water governance tasks are of great importance, while they are designed to resolve manifold water relevant issues, respond appropriately against emergent and exceptive situations and reflect concretely how water laws and regulations are implemented in practical actions [6, 36], Saleth and Dinar 2013). A good way to describe this cooperation is a "process-model", which means describing the workflow shown in a diagram. Therefore, it is crucial to systematically demonstrate different management processes over all involved institutions, which might be a lot for reviewing their competences on practical applications.

Practically, water institutional framework charts are usually drawn to understand water governance structures for different nations or regions, study examples can be seen from [8], Ministry of Water and Environment of Uganda (MWE n.d.), [34] and [44]. Those charts are mostly drawn in various types of 2D layouts and commonly place water relevant organizations at different positions, depending on their hierarchies and organizational functions. On the other hand, water management processes are also commonly depicted in 2D layouts by different forms of flow charts, such examples can be seen from the studies of Bavarian Environment Agency (LfU) (LfU n.d.a, 2010) and [31].

Although the 2D layouts of water institutional frameworks and management processes provide adequate view and understanding of water governance structures and procedures, essential drawbacks also exist. Firstly, since organizations always communicate within the environment of other organizations at parallel and hierarchically levels, namely horizontally and vertically, a 3D relationship thus exists (Grambow et al. 2017). However, water institutional frameworks shown in 2D formats intrinsically could not express such a characteristic. Secondly, water institutional frameworks in 2D layouts cannot fully present all relevant organizations, thus the scale of water governance structures within a specific region could not be viewed in accurate dimensions. Thirdly, regarding the display of water management process charts, usually they are demonstrated statically out of the entire water institutional framework, and provide only limited descriptive information, thus essential background and supportive information is not provided for optimized understanding. Hence, the authors consider that conventional 2D approaches for exhibiting water governance structures and management processes cannot fully express and comprehend the complexity of water institutional structures as well as the spatial dynamic essence of management processes.

The authors have conducted extensive literature reviews to find if there is the existence of previous studies that aimed at developing the 3D water institutional framework (model), and resolved these above-mentioned drawbacks, however, no similar studies have been discovered. In this study, the authors aim at developing two integrative 3D Water Governance Models (3D-WGM) that enable to exhibit all relevant water organizations for the study areas Kunming City, China and Bavaria, Germany, and interactively demonstrate various representative water management processes within the models. In addition, numerous functions are also intended to be incorporated in the 3D-WGMs for facilitating the models' operation.

1.1.1.2 Materials and Methods

Study Areas

With the formation of German research initiative International Partnerships for Sustainable Technologies and Services for Climate Protection and the Environment (CLIENT), which is funded by the German Federal Ministry of Education and Research (BMBF), several projects were launched to support China in solving water resource management challenges in selected areas [9]. Within the CLIENT framework, SINOWATER Project was initiated that aims at advancing Good Water Governance, management and innovative technologies for improving the water quality of Dianchi Lake (falls in Kunming City) and Liao River Basins, where both basins were selected while they are regarded as two of the most important and most heavily polluted water bodies in China. On the other hand, the German experience from Bavarian water resource management and administration was selected in SINOWATER Project to conduct comparative studies for providing water governance strategies and executive suggestions. This study, as a contribution to SINOWATER Project, then chooses the water governance system in Kunming City, China and Bavaria, Germany as study areas.

Kunming City

The first study area where this research was conducted is Kunming City. As a capital city of Yunnan Province, China, Kunming City is located at an altitude of 1,890 m in southwest China, which covers a total area of 21,012 km^2 [27]. The climate of Kunming City is classified as subtropical highland climate (Köppen Cwb) and characterised by warm and humid summers, and cool and dry winters. With one of the mildest climate in China, Kunming City has an annual mean temperature of 15.52 °C and precipitation of 979 mm. The population of Kunming City has reached 6.626 million in which 4.575 million are resided in urbanized regions [45].

A large lake named Dianchi Lake is located at southwest within the jurisdiction of Kunming City. Throughout history, Dianchi Lake has been important for the economy and tourism development as well as many aspects of residents' livelihood in Kunming City. Ranked as the 6th largest freshwater lake in China, Dianchi Lake covers an area of 309.5 km^2 [15, 17]. Over last few decades of rapid economic and population growth, and combined with significant increase of urbanization rate and land use changes, continuous increased pollutant discharges from rural and urban areas have caused severe water quality deterioration and eutrophication of Dianchi Lake, in the meanwhile, ecosystem of Dianchi Lake Basin has been reported that went through a tremendous degradation [15, 27–29].

Since 1980s, the severe water pollution condition in Kunming City, mainly arose from Dianchi Lake Basin, has been widely recognized by national and local governments. Through years of implementing pollution control actions, central and local governments have recognized that fundamental water environment enhancement must be accompanied by strategies of Good Water Governance. Thus, lessons learned regarding water management have been systematically reviewed, and plans in association with water governance were revised and supplemented into existing and future development polices, where the topic of comprehensive understanding and analysis of current water administration systems was also regarded as one of the major future tasks [29].

Bavaria

The second study area is targeted at Bavaria that locates in southeast of Germany. As the largest state in Germany by land area, Bavaria covers a total area of 75,500 km^2. With population amounts to 12,976 million [2], Bavaria is ranked as the second-most-populous state in Germany. While lying in transitional area from Western Europe to Eastern Europe, Bavaria receives climate patterns, which ranges from oceanic climate (Köppen Cfb) to continental climate (Köppen Dfb) (LfU n.d.b). Precipitation also varies greatly both regionally and temporally, on average the total annual precipitation in Bavaria amounts to 933 mm, while in northern Bavaria, it reduces to 768 mm and southern Bavaria reaches 1,058 mm (LfU n.d.c).

Bavaria is a relative water-rich state in Germany, in total 2% of its whole land area, namely 1,446 km^2 is covered by water resources. Approximately, the lengths of all rivers and streams in Bavaria are added up to 100,000 km, where around 9,000 km water bodies are regarded with regional and supra-regional significances [40]. Bavaria also forms parts of three international river basins, namely Danube, Rhine and Elbe, and play important roles in their commissions for protecting river basin environment with joint efforts. With respect to surface water quality in Bavaria, at current stage, the chemical status is overall at good condition (nearly 95% of all the surface water bodies) when not considering the ubiquitous substances, nevertheless, on the other hand, only around 15% of all surface water bodies have reached the ecological good status [26].

History of water management administration in Bavaria dates to 18th Century, until today, steady governmental structures have been formed, which are responsible for sustainable utilization and management of water resources [40]. As members of various task forces (e.g. ARGE Alp, Arge Alpen-Adria) and international river basin commissions (i.e. Rivers Danube, Rhine and Elbe), Bavaria takes further responsibilities of securing healthy water environment in Europe. Furthermore, Bavarian water management administration also extends international collaborations and communications at global scales. Through multiple international collaborative phases, it has been noticeable that each nation or region administrates water resources with its unique governmental and public institutional arrangements, thus it stimulates Bavarian water management administration to systematically and in-depth understand its own water administrative structure as well as the regions with partnerships, such as China, India, Tunisia and Poland.

Establishment of 3D Water Governance Models
Selection of Modelling Software and Programming

Through a wide range of selections, the authors have chosen Unity, a popular interactive 3D rendering engine, to build the 3D-WGMs. We have designed an appropriate description format to encode the water institutional structures in regular Microsoft Excel tables, which are capable of fast and batched data edition, and every time when Unity software launches, it reads all the data into memory and produces the corresponding 3D representation. In this way, meta-data of organizations appeared in the models are separated from programming code to achieve independent editing of organizational data. Hence, each single organization in the 3D-WGMs could be assigned with different sizes in 3D shapes and collectively shown in one single model. Since the 3D-WGMs are required to integrally exhibit abundant of organizations, we have used GPU instancing technique [18], which enabled to render large amount of 3D units in different colours and sizes on one screen in 60 Hz (frame per second, FPS).

Regarding the realization of demonstrating water management processes, we have designed a novel data structure to describe their step-by-step action flows. With this structure, the 3D-WGMs enables to exhibit the action flows as 3D animations in real-time, interactively. In these animations, the models sequentially draw lines connecting two specific units as a representation of actual workflow. When animation encounters branches, a dialog window appears to allow users to choose different options.

Model Construction

With the application of Unity Software, graphical User Interfaces (UI) and various associated applicable functions were developed for both models in Kunming City and Bavaria.

In the first place, each single unit, namely every relevant organization demonstrated in the models, was inserted and assigned with specific 3D shape and colour depending on its type and function. The function of displaying internal structure was incorporated, which allows users to zoom into a specific organization to observe its internal construction at different scales. Subsequently, map images with high resolution, which indicate administrative divisions at different hierarchies were vertically placed with sequences in the models, afterwards, each organization (i.e. the equivalent 3D shape) was placed above the corresponding map for expressing itself as an official organization in the specific area. In addition, each organization was programmed that enables to be edited to provide supplementary literal information (text box placed to display at top-left corner of UI, see model conception Fig. 1.1) for explaining its properties and characteristics.

In the next phase, the function of illustrating the dynamic procedure flows of various representative water management processes both for Kunming City and Bavaria was designed. Every water management process was represented by an interactive button, which is located at top-right of the Model UI (Fig. 1.1). By clicking any of those buttons, the model will initiate to dynamically illustrate this specific water management process step by step, and each action step was realized by utilizing a 3D arrow pointing from one organization to another. In addition, the full literal

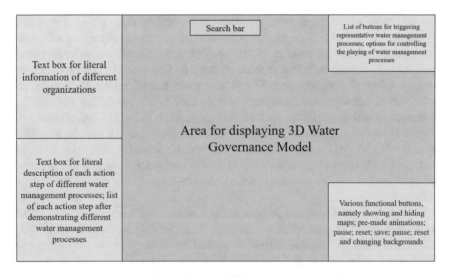

Fig. 1.1 Interface conceptual design of 3D-WGMs

description of each action step was arranged to accompany throughout the illustrative processes, which is placed at the bottom-left of the Model UI (Fig. 1.1). At the end of dynamic illustration of process flows, each relevant action step was arranged to be placed again in order on the Model UI (bottom-left corner, Fig. 1.1) for further understanding, and every step was programmed to link with its corresponding action arrow, which simultaneously shall be highlighted as well.

At later stage of model development, several significant auxiliary functions such as hiding or displaying any 3D unit and map, changing UI background and searching functions (Fig. 1.1), were embedded for facilitating its practical utilization.

Model Information Input

In order to understand and obtain background information to support the establishment of 3D-WGMs, extensive literature reviews both for Kunming City and Bavaria regarding the understanding of water management related institutional arrangements and process descriptions were firstly conducted. During this process, various types of literatures and documents, such as journal articles, official internet documents, reports, statutes, and regulations were retrieved, analyzed and arranged.

Based on essential literature reviews, two rounds of systematic and semi-structured interviews were subsequently conducted both in Kunming City and Bavaria for gathering supplementary information of model establishment. All interviews were executed by the first author and were audio-taped and partially verbatim transcribed. In the first round, according to necessities, interviews were conducted within various representative water relevant organizations in Kunming City and Bavaria. The first-round interviews aimed at clarifying water management structural and institutional constructions. In the second round of interviews, it was designed to explicitly determine the step-by-step description of each selected representative water management process, every single process detail was verified with local water management officers, who are closely involved within those procedures.

1.1.1.3   Results

Construction of 3D Water Governance Models

Graphical User Interface

As results of model development, two individual 3D-WGMs are built for Kunming City and Bavaria respectively. By opening the designated executive file, the User Interface (UI) of Bavarian model presents in the first place. While the model interface integrates both models (Kunming City and Bavaria), it allows users to switch between two models by clicking the appointed button. Since both models are designed with similar UI, thus, in this section, the graphical user interface will be introduced and explained with the example of Bavarian 3D-WGM (Fig. 1.1). The model is composed of several zones, which are numbered and will be elaborated in sequence in the following lines.

1. The first zone represents text boxes, which allows users to enter and edit literal contents for any organization shown in the model. Within the text boxes, users could attach various kinds of documents, such as PDFs, pictures, and video files, for its extended explanation. The text box appears instantly when the mouse cursor is pointed to a specific organization and could be closed manually or replace contents, when the cursor is pointed to another organization.

2. The second zone is reserved for displaying another type of text box, which consists of the description of each action step from any water management process. It is only activated when the process is triggered (see number (5)).

3. The third zone exhibits the main body of the 3D-WGMs. The example in Fig. 1.2 is shown from a specific spatial angle and composed of three parts. In the left part, all the water relevant official organizations are arranged according to their hierarchies and geographical locations on maps. In the right part of this zone, various social organizations, which are involved in water management system, are adjacently displayed above an independent surface. In the middle part, some organizations are placed temporarily, when further arrangement for their positions are necessary.

4. In the fourth zone, namely at the top-middle area of the UI, a search bar is inserted that allows user to search any specific organization, which exists in the model.

5. A list of buttons, which indicates the selected water management processes, is placed in the fifth zone. By clicking any of the buttons, it triggers the dynamic procedure flow of that specific process. During the procedure flow, each

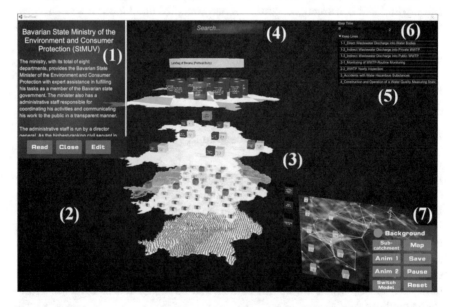

Fig. 1.2 Graphical user interface as the example of Bavarian 3D-WGM at default viewing angle

action step is shown in sequence, which is accompanied by the corresponding explanation shown in the text box at second zone.

6. In the sixth zone, two options are situated for controlling the demonstration of dynamic process flows. The "Step Time" adjusts the speed of flow procedure; on the other hand, during the process flow, "Keep Lines" allows the users to either retain all the action lines on the UI or enable the action line to disappear before the display of the subsequent action line.

7. In the seventh zone, several interactive checkbox and buttons are designed, which allows general or additional operations and will be explained in succession. Checkbox "Background": it provides users the option to enable the display of the background panoramic picture or uncheck to hide the background. Buttons "Anim 1" & "Anim 2": once activated, these two pre-made animations will automatically rotate and zoom through the main body of the 3D-WGMs from different angles, which gives comprehensive visual tour along and among the 3D constructions. Button "Switch Model": this button simply allows users to switch between the 3D-WGMs of Kunming City and Bavaria. Button "Sub-catchment": analogue to a "spotlight" function, when it is required to selectively exhibit a specific group of organizations, and in the meanwhile hide other irrelevant organizations, this button could help to realize the requirement by appropriate pre-settings. In addition, more similar buttons could be added based on further necessities. Button "Map": by clicking this button, the window, which contains a list of all displayed maps in the 3D constructions, will present. At this moment, users have the possibility to separately keep or hide the desired maps. Button "Save": this button should be utilized when editing the position of a single or several organizations. After the editing operations, the "Save" button should be clicked to save the changes. Button "Pause": during the procedure of displaying water management processes or pre-made animations, pause button could be used to suspend these actions. Button "Reset": by clicking this button, the model will automatically bring the users back to view the 3D constructions at default angles.

8. The overall operation of the 3D-WGMs is realized by the combination of mouse and keyboard. It has been programmed that the two main buttons of the mouse are responsible for adjusting the viewing distance horizontally or vertically and changing the viewing angles at all directions. In addition, the mouse wheel button could be scrolled to zoom the 3D constructions at any depth. Thus, the 3D-WGMs could be observed at any level of details within the limitless 3D space (see Appendix A and Appendix B).

Individual Organization and Internal Structure

Each single organization forms as the basic unit of the 3D-WGMs and is assigned with its specific 3D shape. In general, most of the common organizations are designed with cube or cuboid shapes, in addition, some special organizations are built with cylinders and sphere shapes (Fig. 1.1, 1.2, 1.3). The size and proportion of each organization refer to its best fit with the map, which lies underneath.

Fig. 1.3 Examples of individual organizations, which are assigned with cubic, cuboid, cylinder and sphere shapes

Table 1.1 The attributes of colors and corresponding indicated functions for the establishment of organizational shapes

Colour name	Indicated function
Red	Chinese Communist Party organizations
Yellow	General political bodies
Blue/Light Blue	General administrative bodies (two types of variations)
Green	Water supply enterprises
Brown	Wastewater treatment enterprises
Dark green	Hybrid water supply and wastewater treatment enterprises
Light Orange	Other types of enterprises
Black	Organizations with temporary positions (need further adjustments)
Pink	Technical institutions
White	Social bodies

While each organization is established with its major purposes, it is thus necessary to assign color attribute to indicate its type of function. Several different typical colors are used to distinguish the most common organizational functions and is shown in Table 1.1.

Due to the models' 3D spatial characteristics, the internal structures could also be built within the 3D shape of any specific organization. The construction of internal structures could be extended to various scales, which is not only limited to two level of details. If one organization's internal structure has been designed, one simple click will bring user into the organization's internal view and conduct further observations. Figure 1.1, 1.2,1.3,1.4 gives an example of one organization's internal construction in three levels.

Application of 3D Water Governance Models in Kunming City and Bavaria
3D Water Governance Model of Kunming City

As illustrated in Fig. 1.5, the front view of the Kunming City WGM illustrates the Kunming water governance system in five official hierarchies, namely Yunnan provincial level, Kunming City level, county (district) level, township (sub-district) level and village (residential community) level. Although the model focuses on

Fig. 1.4 Regional Water Management Office Deggendorf (WWA-Deggendorf) in Bavarian 3D-WGM as an example to demonstrate the construction of organizations' internal structures. (a) represents the WWA-Deggendorf itself; (b) represents the first level internal structures; (c) represents the internal units of one specific department in WWA-Deggendorf

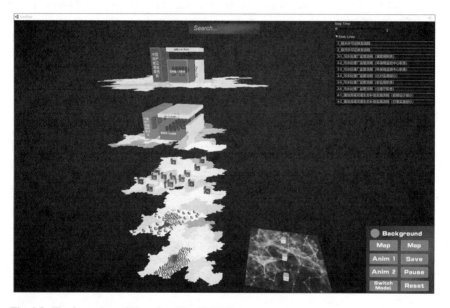

Fig. 1.5 The front view of Kunming City 3D-WGM

exhibiting the water governance model in Kunming City, the provincial organizations also play significant roles in local water administrations, thus they have been also incorporated in the model. In general, four major types of governmental groups, namely Party Committees, governments, People' Congresses (PC) and People's Political Consultative Conference (People's PCC), are widely and similarly established at each jurisdictional local region. While at lower hierarchies exceptions exist, where PCs are not established at village level and People's PCCs do not appear at both township and village levels. Different governmental groups play respective roles in governing all socioeconomic aspects, which include water resources.

In ordinary, local government is the highest administrative body in each region, which plays fundamental roles in administrating various socio-economic affairs.

Local government should execute the resolutions from its corresponding PC and implement orders from higher administrative bodies, in the meanwhile, it also leads the administrative work of lower hierarchies within its jurisdictions. Local PC represents the highest power authority within its territory. Its major functions are formulating and promulgating necessary local regulations; ensuring the implementation of constitution, manifold laws and regulations; monitoring and examining the performances of its corresponding government etc. Local People's PPC serves as political consultative body to its related PC and Party Committee in the same region. Regarding the function and power of local Party Committee, it plays central and comprehensive roles in leading and coordinating the overall performance of local governmental groups. A local Party Committee is the prime organization in a given region to discuss and adopt the resolutions on major issues and policies, in the meanwhile, it is governed by a top-down approach, which is responsible to the Party Committee at the next higher level.

As illustrated in Fig. 1.5, most of the organizations related to water administration in Kunming City are displayed in 3D-WGM, which are based on their hierarchical and geographical positions. From higher to lower hierarchies, each level's major functions regarding water administration are described in the following lines:

Provincial Level

Governmental organizations in provincial level involve in Kunming City water administration by various approaches. Firstly, numerous significant water related local statues and regulations are promulgated at provincial level, and essential central policies are interpreted ahead in provincial level, which could further guide the implementations at lower hierarchies. Secondly, provincial government examines the water administrative performances at city level. In addition, some further core involvement mechanisms from provincial level also exist, such as financing broad range of projects and campaigns; setting specific targets, which is required to be achieved by lower hierarchies; and issue permissions for actions that have higher water environmental impacts.

City Level

City governmental organizations play central roles in administrating water resources in Kunming, especially for the urbanized regions. In the first place, amounts of water relevant regulations are issued by Kunming City government and PC, which are formulated for ensuring their applicability in Kunming. Secondly, essential concrete water administrative tasks are performed by various bureaus and intuitions (arranged as internal construction of Kunming City government in its 3D-WGM) in Kunming City government, those tasks cover broad range of water administration, such as issuing permissions for water extraction and wastewater discharge, conducting water quality monitoring, legal controls against unauthorized activities, and flood risk prevention. Furthermore, Kunming City also coordinates and supervises governmental organizations in county level for implementing water related socio-economic development plans and various local policies.

County Level
17 county level groups of governmental organizations are demonstrated in the 3D-WGM. In general, county levels are also responsible for executing routine water administrative tasks within their territories, while those tasks are either assigned by city level or required by relevant regulations. As described in city level, county levels are also responsible for fulfilling the assigned tasks of water planning and policies.

Township and Village Level
68 and 214 groups of respective township and village governmental organizations (within Dianchi Lake Basin) are exhibited in the 3D-WGM. Regularly, governmental organizations from both levels are not actively involved within significant water administration processes. Their main responsibilities usually consist of ensuring the functioning of water conservancy facilities, performing flood and drought risk controls and executing necessarily assigned duties from higher levels.

3D Water Governance Model of Bavaria
As Fig. 1.2 captures the side view of Bavarian 3D water governance constructions. In total, four governmental hierarchies, namely ministry level, administrative district level (Regierungsbezirke), county or county-free city level (Landkreis or Kreisfreie Stadt) and municipality level (Gemeinde) constitute the foundation of governmental administration systems in Bavaria. Within each administrative region, governmental organizations are composed of political bodies and administrative authorities. Regarding the functionalities, each political body is in general the main organ of its administrative region, it decides on all the significant regional affairs and establish principles for the performance for its corresponding administrative authority. The highest political body (Landtag of Bavaria) in Bavaria has the authority to issue laws, while the other political bodies in lower hierarchies could only promulgate statutes or legal provisions, which are below laws and formulated based on the regional conditions. On the other hand, administrative authorities are responsible for executing routine administrative tasks and resolutions from their parliaments or councils. In the next lines, each governmental level will be described regarding their major functions for Bavarian water administration.

Ministry Level
Exhibited in the 3D-WGM as one of the 11 state ministries in Bavaria, the Bavarian State Ministry of the Environment and Consumer Protection (StMUV) functions as the highest state authority for water resource management. Its major responsibilities are firstly to operate as the sole legislative body that develops and implements various water relevant state laws and provisions. Secondly, StMUV provides guidance and monitors the overarching water administrative procedures in whole Bavaria, it goes through most essential decisions on water administration, issue permissions for substantial proposals and set management strategies for significant water environmental events. In addition, it also plays roles in national and international water resource and river basin commissions for transboundary cooperation.

Subordinate to StMUV, LfU is the central public authority for managing environment related scientific and technical issues in Bavaria. It specializes in gathering and evaluating all aspects of water relevant data and further develops objectives, strategies, and plans for water environment protection. As advisory agency, LfU provides professional advices for public authorities and water management institutions, such as StMUV, Regional State Offices for Water Management (WWAs), and county water management offices. Equally important, it also bridges publics by providing raw or compiled water data, research or monitoring reports and forecast or warning service against hazards.

In addition, 17 WWAs are displayed in Bavarian 3D-WGM as special "expert" authorities in between district and county levels. WWAs are also established as subordinate technical public authorities to StMUV, and evenly distributed in Bavaria for executing local water tasks. They are by law "official experts" which means they are responsible to offer the know-how of the integrated water management on the regional level. In that meaning, they are independent official experts. In that role, the WWAs provide essential guidance and technical advices for county administrative authorities (which in legal meanings are responsible for the water management what means e.g. permitting water extraction and wastewater discharge applications, allowance for river construction and so on) and communities in terms of establishing and managing sanitation, water supply and wastewater treatment facilities. On the other hand, they are directly responsible for performing some delegated tasks from state, such as developing (e.g. flood protections, riparian restoration) and maintaining important state water courses, constructing and maintaining water infrastructure facilities including the associated properties, but especially the complete water monitoring system as chemical and biological water qualities, hydrology, etc.

Administrative District Level

Classified as the upper public water authorities (exhibited as second hierarchy in Bavarian 3D-WGM), seven district governments in Bavaria are mainly responsible for supervising and coordinating various water relevant stakeholders within their territories to realize a common solution or compromise, where issues could not be fully accomplished by local administrative authorities.

County or County-Free City Level

Overall, governments of 71 counties (Landratsamt) and 25 county-free cities (Kreisfreie Städte) are displayed at the third administrative level in the Bavarian 3D-WGM. The water rights units (specialist departments) situated in governments at county level are the offices responsible for ensuring water relevant laws and statues to be universally obeyed within respective county region. They are regarded as official authorities for water legal matters, thus, all water activities, which are required to be permitted or certified, shall be firstly examined and approved by the water rights units. In addition, they also make administrative decisions regarding illegal activities, water hazardous accidents, etc.

Community Level
Ranked as the smallest administrative divisions in Bavaria, in total 2,031 communities are placed at the lowest level of Bavarian 3D-WGM. With the sovereign rights, communities are empowered to manage several water activities on their own responsibility, which include providing services for local water supply and wastewater treatment and managing the 3rd order water courses (1st and 2nd order water courses are regarded as important water bodies, which are managed by the states).

Water Management Processes
One of the major applications of the 3D-WGMs is dynamically and stepwise demonstrating diverse water management processes. Since each water administrative system comprises numerous water management processes from different aspects, such as monitoring, water supply, wastewater treatment, and hydraulic construction, and not every process could be investigated in detail within this study, thus, four representative water management processes were chosen both for Kunming City and Bavaria, which are listed in Table 1.2. The table also indicates that some processes could contain sub-processes, depending on the actual local situations. This section will only select and elaborate one comparable process from each study area, namely issuing wastewater discharge permission into public sewage system in Kunming City and issuing permission for indirect wastewater discharge into public wastewater treatment plants in Bavaria (sub-project), to demonstrate as examples for introducing the displaying mechanism of water management processes in the 3D-WGMs.

Firstly, the management process issuing wastewater discharge permission into public sewage system in Kunming City aims at ensuring the wastewater discharge activities from different entities, such as industrial enterprises, individual businesses, and hospitals in Kunming urbanized region, to comply with the connectivity requirements with sewage system. This permission mainly examines that whether the local wastewater dischargers could safely transport their wastewater into public wastewater treatment plants. With essential keywords, Fig. 1.6a exhibits the entire flow diagram of this water management process, while the detailed stepwise description is attached in the Table 1.3 (in sequence of numbers shown in Fig. 1.6b). On the other hand, Fig. 1.6b displays the ending view of the same process, which is dynamically demonstrated in its 3D-WGM.

Table 1.2 Selected representative water management processes in Kunming City and Bavaria for the demonstration in 3D-WGMs. *N/A: Not applicable.

Region	Process number	Sub-process number	Name of process
Kunming City	1	N/A*	Process of issuing wastewater discharge permission into public sewage system
	2	N/A	Process of issuing permission for general pollutant discharge
	3	N/A	Process of monitoring wastewater treatment plants

(continued)

Table 1.2 (continued)

Region	Process number	Sub-process number	Name of process
	4	N/A	Process of executing river ecological compensation in Dianchi Catchment
Bavaria	1	1–1	Process of issuing permission for direct wastewater discharge into water bodies
	1	1–2	Process of issuing permission for indirect wastewater discharge into private wastewater treatment plants
	1	1–3	Process of issuing permission for indirect wastewater discharge into public wastewater treatment plant
	2	2–1	Process of monitoring wastewater treatment plants (routine monitoring)
	2	2–2	Process of wastewater treatment plant yearly inspection
	3	N/A	Process of dealing with accidents by water hazardous substances
	4	N/A	Process of constructing and operating a water quality measuring station

* Kunming DAB: Kunming Dianchi Administrative Bureau.

Secondly, the wastewater related permissions in Bavaria generally comprise three sub-processes (see Table 1.2). Regarding the sub-process issuing permission for indirect wastewater discharge into public wastewater treatment plants in Bavaria, it is responsible for ensuring the wastewater discharge activities are in compliance with the local corresponding sewage relevant regulations. Similarly, the entire flow diagram of this process is shown by Fig. 1.7a and its detailed stepwise descriptions are listed in Table 1.4 (in sequence of numbers shown in Fig. 1.7a). Additionally, Fig. 1.7b exhibits the spatial process flow, which is illustrated in the Bavarian 3D-WGM.

1.1.1.4   Discussion

Strengths and Application Value of 3D Water Governance Models
The development of 3D-WGMs has provided new approaches for comprehending local water governance systems. Several significant advantages of its application are discussed in the following lines.

The first strength of the 3D-WGMs indicates its integrative capability, which enables to display each single organization that is involved in local water governance system. Thus, this feature could give the model users the direct impressions of the system's dimension, regarding its layer of hierarchies, number of involved

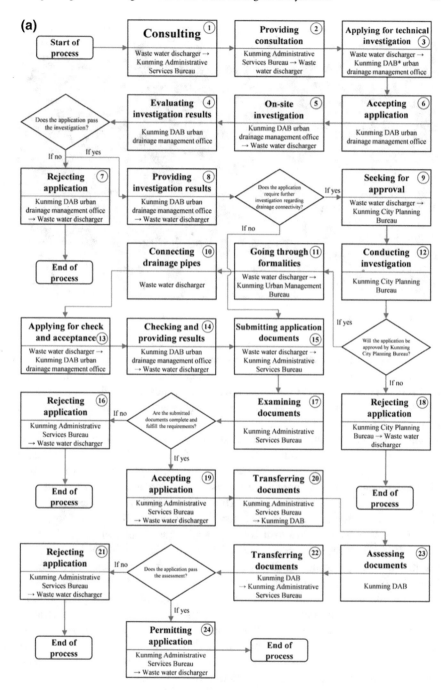

Fig. 1.6 Illustrations of water management process of issuing wastewater discharge permission into public sewage system in Kunming City. (a) 2D flow diagram of the entire processes; (b) exhibition of the process by Kunming City 3D-WGM, where all the options are answered with "yes"

Fig. 1.6 (continued)

organizations and corresponding spatial location of each involved organization. On the other hand, each organization is built with independent 3D shape, which enables the users not only to observe the water governance system as a holistic structure, but also to zoom into the internal construction of any specific organization for further understandings. Hence, this feature allows to comprehend different water governance systems from both macro and micro perspectives. Comparing to the conventional 2D displays of the water institutional framework from several previous alike studies [8], MWE n.d.; [34, 44], the 3D-WGMs manifest their optimization of integrating different dimensional and complete structural information into one single framework.

The second advantage of using the 3D-WGM is based on its 3D and interactive characteristics. On one hand, the 3D essence of the models allows users to observe the water governance systems from different angles and depths, which is adjusted to the users' preferences and requirements. On the other hand, with the programming results from Unity engine, the 3D-WGMs enable users to conduct various interactive operations, such as viewing institutional background information of different organizations, searching for specific organizations, hiding or displaying preferred institutions, generating observational animations, etc. Therefore, those interactive operations permit users to comprehend water governance systems based on individual requirements and focus on specific details. In comparison with previously mentioned studies, those institutional frameworks could only provide knowledge from one or several aspects, but not able to provide further interactive details from users' expectations.

The third significant feature of 3D-WGMs emphasizes that it combines the dynamic display of various water management processes within the models. Again,

Table 1.3 Detailed step-wise description of the water management process issuing wastewater discharge permission into public sewage system in Kunming City (based on Fig. 1.6)

Action step No.	Action step content
1	At the first place, the wastewater discharger should consult the Kunming Administrative Service Bureau for general aspects of wastewater discharge issues, knowing about its service domain, guidelines and forms for applying permissions
2	Kunming Administrative Service Bureau will then provide consultations to wastewater discharger for answering common questions regarding preparing documents and application procedures
3	According to standard procedure, wastewater discharger applies for a required technical investigation regarding drainage connectivity, which is conducted by the Kunming DAB urban drainage management office
4	After technical investigation, Kunming DAB urban drainage management office organizes experts to evaluate the investigation results, this step examines whether the aspects, such as connection of pipelines, separation of storm water and wastewater, utilization of recycled water etc., are in compliance with the city regulations and requirements
5	Kunming DAB urban drainage management office then starts to assign experts to conduct the technical investigation on-site
6	Kunming DAB urban drainage management office officially accepts the application for technical investigation
7	If the application does not fulfil the requirement of the investigation, Kunming DAB urban drainage management office will reject the application from wastewater discharger
8	If the application passes the investigation, Kunming DAB urban drainage management office will provide the investigation report to wastewater discharger for further arrangements
9	If the application needs to apply further investigations regarding drainage connectivity (for new or extended projects), the wastewater discharger then need to seek for approvals by Kunming City Planning Bureau, who is responsible for granting permission of road and drainage construction
10	Once approved, the wastewater discharger can start the construction to connect its drainage system
11	If the application is approved by the Kunming City Planning Bureau, wastewater discharger will then go through formalities to obtain the approvals
12	When receiving the application, Kunming City Planning Bureau will conduct its investigation regarding construction issues
13	Once the construction is finished, the wastewater discharger should then apply by Kunming DAB urban drainage management office again for checking the construction results and seek for its acceptance

(continued)

Table 1.3 (continued)

Action step No.	Action step content
14	Kunming DAB urban drainage management office further examines the construction quality and gives comments back to wastewater discharger, here inappropriate details should be changed and amended by wastewater discharger
15	At this step, the wastewater discharger officially submits its application documents to Kunming Administrative Service Bureau for standard procedure of applying the permission. Here complete documents should be submitted, which include application form, environmental impact assessment report, business licenses, reports and assessment from previous steps, etc
16	If the documents submitted cannot meet the requirements or not complete, the Kunming Administrative Service Bureau can reject the application, or ask for its re-submission
17	Once the Kunming Administrative Service Bureau receives the application documents, it will start to examine the completeness and effectiveness of the required documents
18	If the application is not approved Kunming City Planning Bureau, it can reject the application from wastewater discharger or ask for amendments and re-submission
19	If the received documents fulfil the requirements, the Kunming Administrative Service Bureau will accept the application and give confirmation documents to wastewater discharger
20	Kunming Administrative Service Bureau further transfers all the documents to Kunming DAB for its final assessment
21	If the application is not approved, then Kunming Administrative Service Bureau will inform the wastewater discharger about the decision of rejection
22	After assessment, Kunming DAB will make the final decision and deliver all the necessary documents back to Kunming Administrative Service Bureau
23	When Kunming DAB receivers all the documents, it will conduct the final assessment, which makes general examination by considering all relevant aspects
24	If the application has been approved by the final assessment, then the Kunming Administrative Service Bureau will issue the permission licence to the wastewater discharger, and it can officially discharge wastewater according to the requirement of permission

with application of the interactive functions from Unity engine, abundant water management processes could be built for each 3D-WGMs, which indicate how each single water management task is resolved by the communication and interactions among different relevant organizations. The display of each management process is designed with several key practical features, which are embedded within the models. Firstly, when any specific management process is triggered, the model will automatically hide the organizations, which are irrelevant to the process, this feature helps users to distinguish which are the involved organizations and where are they located.

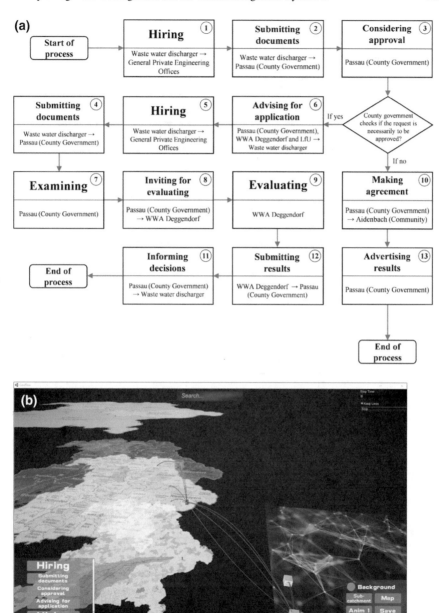

Fig. 1.7 Illustrations of water management process of issuing permission for indirect wastewater discharge into public wastewater treatment plants in Bavaria. Here the example is made by including Passau County Government, WWA Deggendorf and Aidenbach Community. (a) 2D flow diagram of the entire processes; (b) exhibition of the process by Bavarian 3D-WGM, where all the options are answered with "yes"

Table 1.4 Detailed stepwise description of the water management process issuing permission for indirect wastewater discharge into public wastewater treatment plants in Bavaria (based on Fig. 1.7)

Action step no.	Action step content
1	In the first place, the wastewater discharge applicant will need to hire engineering office for helping to fill the necessary documents, which indicates the general planning for its discharge activity
2	After the preparation of the general planning description together with the engineering office, the wastewater discharger submits the documents to the corresponding county government, here is Passau (County Government)
3	After receiving the descriptive documents, Passau (County Government) needs to examine if the application should to be approved according to the relevant regulation, namely the Waste Water Ordinance
4	Subsequently, the wastewater discharger submits its official application documents to Passau (County Government)
5	After the application conference, ideally the wastewater discharger will hire the engineering office again to prepare the official discharge application documents
6	If the approval is needed, then Passau (County Government) will initiate an application conference, there it advises the wastewater discharger how to fill the official application documents and inform about the contents and the extend of the documents as well. When necessary, the Passau (County Government) will also invite LfU and WWA Deggendorf to participate the application conference for consultation
7	After receiving the application documents, then the Passau (County Government) checks their completeness and the sufficient effectiveness. If Passau (County Government) finds that there are missing documents, then it will demand the applicants to re-submit the corresponding documents. When necessary, LfU and WWA Deggendorf should also get involved for helping Passau (County Government) to check the received application documents
8	In the next step, Passau (County Government) sends all application documents to WWA Deggendorf and invites the experts from WWA Deggendorf to conduct professional assessment
9	Afterwards, the official WWA Deggendorf experts receive the documents, they should start to evaluate the documents and formulate their expert opinions. This is considered as the most significant step in the entire process, while their opinions could significantly influence the decision of Passau (County Government). The general task of the official experts is to investigate the facts and the associated effects of the applications. According to the requirement of Administrative Procedure for the Implementation of Water Law, the expert opinions should be created within a period of three months. The experts should present the expected impacts of the application on rights and the related legal interests of the involved parties, also including the state. During the process, WWA Deggendorf experts should consider all the relevant laws and regulations to formulate their proposals. The experts should also formulate their opinions on their own responsibilities and not get other instructions. In order to create this proposal, the experts need observe some guidelines for their work, for example the expert report model and the information sheet from LfU. For their participation in the processes, the official experts should be compensated for their costs, as far as they could be charged from the wastewater discharger

(continued)

Table 1.4   (continued)

Action step no.	Action step content
10	If no approval is needed, then Passau (County Government) just needs to have an agreement with the local drainage rules by the local municipality, which will allow the applicant to discharge into the wastewater treatment plant in that community, here is Aidenbach (Community)
11	Depending on the proposal from WWA Deggendorf, the Passau (County Government) makes final legal decisions, they should inform the wastewater discharger if the permission is granted or not. This decision is then not open for discussion and only can be challenged later by the administrative court
12	At the end of the evaluation, the official experts from WWA Deggendorf should submit their final proposal to Passau (County Government) which could be used as the basis for its final decision on the application. In this proposal, the experts should indicate that if the impairment of the application to the well-being of the public is not a general concern and applies to the content and subsidiary regulations. On the other hand, if not, the experts should indicate that the welfare of the publics is not secured, although the existence of some secondary appropriate measures, then they suggest rejecting to give the permission for discharge according to Federal Water Act
13	The Passau (County Government) should advertise publicly the results

Secondly, each action step of the processes is exhibited with logical sequences and accompanied by full literal explanations displayed parallel with the process flow. Thus, it facilitates that users could go through a stepwise explanation of any specific process and informed by rich description of each single step. It was also designed that the dynamic process could be paused at any moment. Thirdly, when the management processes are completely shown, all the previous action steps will be listed again, and users could at this moment replay any of the steps to conduct in-depth learnings, and in the meanwhile, this selected step will also be highlighted in the model. Therefore, this function provides the advantages to fully understand a process by post-learnings. While management processes are key elements of understanding the administrative mechanisms, organizational relationships and information flow [6, 36], Saleth and Dinar 2013), the integration of the management processes in the 3D-WGMs has facilitated to understand the connections among different organizations. In many circumstances, different water management processes are normally depicted by various kinds of 2D flowcharts, such as the examples that are existed by LfU (n.d.a, 2010) and [31]. In comparison, the most significant benefits of applying the management processes within the 3D-WGMs are that it is dynamically shown and integrates numerous management processes within one single model.

The application value of 3D-WGMs could be potentially seen in several fields. In the first place, water related governmental organizations at different hierarchies could adopt the 3D-WGMs for understanding own water governance systems from different perspectives. New ideas or improvement strategies could emerge, when governance structures are comprehensively examined. In addition, governmental organization could also utilize the 3D-WGMs as educative tool for training new employees and as guidance tool for facilitating employees' routine work. In the second place, 3D-WGMs could also be provided to general publics for guiding citizens and different public groups to handle water related activities, such as applying permission for wastewater discharge, the requirements to construct and maintain small wastewater treatment facilities, etc. On the other hand, by providing the 3D-WGMs, it also indicates that local water administration respects the public relationships and governs water with more transparency. Finally, general researches could also apply the 3D-WGMs as auxiliary tool for conducting comparative studies of water governance experience, further lessons could be learned when multiple 3D-WGMs are parallel compared and investigated.

Limits of 3D Water Governance Models

The 3D-WGMs have been developed with various benefits that are possible for potential practical applications, nevertheless, in this section some of their limited aspects will also be addressed for an objective discussion.

The first limit of 3D-WGMs arises from their complexity. While the models are developed analogous as a software, it instinctively would require a longer learning curve to get familiar with their mode of operations, design of UI, functionalities of various interactive fields, etc. In comparison with conventional 2D administrative frameworks or organizational charts, 3D-WGMs appear to be less forthright and more complicated, however, after investing adequate amount of learning time, the 3D-WGMs can be mastered with their full range of capabilities. Secondly, the maps utilized in the 3D-WGMs are currently represented by raster graphics, where geographic information could not be stored by each single 3D shape in the model. Thus, being lack of geographic attributes, limits exist that the models could not be operated with more realistic simulations. However, although without added information, the maps that are made from raster graphics could adequately fulfil the purposes of the establishment of 3D-WGMs. Thirdly, the 3D-WGM is currently at the ending phase of its development. While concerning to advance its practical applications in local water governance systems, further effort should be devoted to promoting its capabilities and persuading decision makers to integrate them into their management systems.

Summary and Outlooks

In this study, a novel and integrative 3D Water Governance Model (3D-WGM) was developed for demonstrating a complete local water governance system and displaying representative dynamic water management processes. Throughout this article, the model building tool was firstly introduced, and the model construction concept was explained. Subsequently, the functionalities of the model were elaborated, and two 3D-WGMs developed for Kunming City and Bavaria were in detail demonstrated for indicating the model's practical application. At the end, the strengths, application value and limits of 3D-WGMs were respectively discussed.

Regarding future work, several aspects could be considered to improve the functionality and extend the applicability of the 3D-WGMs. In the first place, more effective features could be integrated into the models, such as enabling vector-based maps, allowing assigning different attributes to organizations or action step of water management processes, and visualizing accountabilities among organizations. In the second place, more example studies can be conducted beyond local water management systems, such as at national and continental levels. In addition, with the facilitation of 3D-WGMs, future studies can comparatively articulate the water governance characteristics between Kunming City and Bavaria, and the 3D-WGMs could be accordingly improved to optimize the water governance system in study areas.

1.1.2   Water Governance Structural Characteristics of Kunming City and Bavaria and Their Comparison

1.1.2.1   Introduction

The remarkable growth in China's economy and population over the past three decades has caused tremendous degradation to the country's water environment. Since 1996, Chinese governments in different levels have begun to take actions to treat polluted water bodies and enhance overall water governance performances, those activities are closely embedded in the corresponding Five-Year-Plan and Major Water Program [5], Ministry of Ecology and Environment of China n.d.).

In order to facilitate Chinese Major Water Program, Federal Ministry of Education and Research of Germany (BMBF) and Ministry of Science and Technology of the People's Republic of China (MoST) has jointly initiated SINOWATER Project which aims at promoting Good Water Governance, management and innovative technologies for improving the water quality of two Chinese prominent and heavily degraded water bodies - Liao River Basin and Dianchi Lake Basin [9].

Designed as a contribution to water governance perspective of SINOWATER Project, this study selects Kunming City, China and Bavaria, Germany as study areas to systematically understand their respective water governance essential structural characteristics. In this study, qualitative interview analytical approach will be the first time parallel employed among water administrative systems in Kunming City and Bavaria. The results of the study will provide in-depth elaborations and comparisons of crucial water governance structural features from the study areas, which could further bring inspirations for the optimization of both water governance systems.

1.1.2.2   Materials and Methods

Participants and Data Collection
The data collection process consists of two phases of qualitative interviews, which both were conducted locally in our study areas – Kunming City and Bavaria.

The first phase of qualitive interviews was designed to collect data for systematically understanding the local water management fundamental structures and the major responsibilities of the interviewed organizations. For both study areas, semi-structured interview type was determined as suitable approach for the research, and a purposeful snowball sampling strategy [30] was used to select most suitable interviewees. During the interview procedure, pre-designed interview guide was always employed for facilitating the entire process.

For Kunming City, the authors firstly established contacts with officers from Kunming Academy of Environmental Sciences (KAES), after negotiations and discussions, they have been fully informed with the purposes of the qualitative interviews and recommend 12 most representative water relevant organizations (Table 1.5) for conducting interviews. In addition, official letters were transmitted from KAES to those organizations for requesting cooperation. In the end, 7 individual interviews and 9 focus group interviews were conducted within 12 official organizations in 2016, in total, 38 people participated the whole interview procedure. The durations of individual interviews lasted around 1 h, and the lengths of focus group interviews varied from 2 to 4 h.

As to Bavaria, the authors consulted several officers from Bavarian water administrative system, who are also members of SINOWATER Project, they recommended 4 representative organizations (Table 1.6) for executing qualitative interviews and further helped to select suitable interviewees and arrange appointments. In the end, 5

Table 1.5 List of interviewed organizations in Kunming City

Kunming City interviewed organizations (English name)	Kunming City interviewed organizations (Chinese name)
Kunming Academy of Environmental Sciences (KAES)	昆明市环境科学研究院
Kunming Dianchi Administration Bureau (KDAB)	昆明市滇池管理局
Kunming Environmental Protection Bureau (KEPB)	昆明市环境保护局
Kunming Appraisal Center for Environmental Engineering (KMACEE)	昆明市环境工程评估中心
Kunming Dianchi Ecology Institute (KDEI)	昆明市滇池生态研究所
Kunming Water Resources Bureau (KWRB)	昆明市水务局
Kunming Dianchi Investment Co. Ltd (KDI)	昆明滇池投资有限责任公司总部
Kunming Dianchi Water Treatment Co. Ltd (KDWT)	昆明市滇池水务股份有限公司
Kunming Dianchi Treatment and Development Co. Ltd (KDTD)	昆明滇池湖泊治理开发有限公司
Kunming Dianchi Drainage Facilities Management Company (KDDFMC)	昆明排水设施管理有限责任公司
Kunming Environmental Monitoring Center (KEMC)	昆明市环境监测中心
Kunming Water Supply Group Co. Ltd (KWSG)	昆明自来水集团有限公司

Table 1.6 List of interviewed organizations in Bavaria

Bavaria interviewed organizations (English name)	Bavaria interviewed organizations (German name)
Bavarian State Ministry of the Environment and Consumer Protection (StMUV)	Bayerisches Staatsministerium für Umwelt und Verbraucherschutz (StMUV)
Bavarian Environment Agency (LfU)	Bayerisches Landesamt für Umwelt (LfU)
District Government of Oberbayern (ROB)	Regierung von Oberbayern (ROB)
Regional State Offices for Water Management – Munich (WWA-M)	Wasserwirtschaftsamt München (WWA-M)

individual interviews and one focus group interview were conducted within 4 official organizations in 2016, and in total 7 people participated the overall interview process. The reason why number of interviewees is much fewer in Bavaria than Kunming City is that much previous literature reviews have helped to identify significant amount of fundamental interview questions. The durations of individual interviews varied from 45 min to 1 h and the focus group interview lasted 1 h. For both study areas, all participants' information was anonymized. The authors took field notes and digitally recorded the whole interview processes, afterwards all interviews were fully transcribed in Chinese, English and German, depending on the languages which were used during conversations.

Since understanding the characteristics of representative water management processes within our study areas is one of our major research targets (see Sect. 1.2.3.1.3), thus the second phase interviews were conducted both in Kunming City and Bavaria. During this process, as listed in Table 1.2, 4 representative water management processes were selected in each study area for in-depth investigation. The authors firstly gathered available documents about those management processes, then with the help of the previously contacted persons, corresponding experts and officers who are indeed involved within those management processes were identified, and further contacts were made to request for conducting interviews. Finally, adequate specific and more structured interviews were made with those experts and officers, and each single detail of those selected management processes was discussed and clarified. The lengths of those interviews varied from 1 to 2 h. The information of those interviewees was consistently kept anonymous. During the interviews, the authors took field notes and organized them into detailed stepwise process descriptions, which are demonstrated alike as the examples Figs. 1.6 and 1.7 have shown.

Data Analysis

The authors read the transcripts and field notes numerous times, memos were written about their general and specific impressions. Then the data was analyzed according to the grounded theory approach, namely using constant comparative method to code it and find patterns and potential theories throughout the overall coding procedure [30]. Following the guidelines, the first step of the analysis was open coding, where data was examined line by line to identify the more detailed and specific characteristics about the water governance structures of Kunming City and Bavaria. In this step, authors got close to data and always allowed new ideas to emerge for formulating codes, and codes were constantly compared. In the meanwhile, authors constantly kept writing memos in order to record the emergence and transformation of various codes. In the next step, method of focused coding was adopted to re-examine and sort the initial codes (around 350 in total) in order to advance them into categories, which were more abstract and could theoretically tell the common essence of the subsumed codes. In the end, the most significant categories were selected for further elaboration in the results and discussion section. Throughout the analysis path, qualitative data software – Nvivo 12 was used to aid data analysis.

1.1.2.3 Results and Discussion

Water Governance Structural Characteristics of Kunming City
Extensive and Sophisticated Water Governance Constructions
As has been introduced already in Sect. 1.2.3.1.3, five official administrative hierar-chies with their different types of governmental organizations play roles in managing general water issues of Kunming City, where their basic functions on water gover-nance have been shortly explained. Within those organizations, the interview study found that city government has the most essential force in administrating and managing concrete water management tasks, and the governmental information portal has shown that Kunming City government consists of numerous bureaus, offices and commissions, which jointly manage water resources (Kunming City Government n.d., Fig. 1.8). Within those organizations, the interview analysis recog-

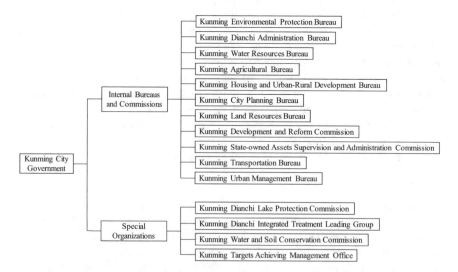

Fig. 1.8 Water governance engaged internal organizations of Kunming City government

nized that KEPB, KDAB and KWRB could be regarded as the three most important water related city bureaus, they are established at parallel administrative level and function with different prime concerns. KEPB is mainly responsible for topics about water quality assurance, water pollution treatment and reduction as well as the corresponding legal and policy issues. KDAB is established as a bureau for managing, monitoring, legally controlling and coordinating Dianchi Lake Basin management and treatment issues. KWRB is then the bureau takes responsibility for general water resource management in Kunming City, where hydraulic, water supply, flood and drought risk issues are at focuses. Besides these three bureaus, there are also other parallel ones, which are partially responsible for water issues, for instances, Kunming Land Resource Bureau is responsible for partially managing groundwater that has heat values; Kunming Agriculture Bureau targets at water saving matters at agriculture sector; Kunming Transportation Bureau has rights to manage related issues on waterways, such as legal controls, traffic safety assurance and pollution emergency treatment, etc. Thus, we could relate that not only the central government consists 9 dragons that jointly manage water resources [43], but also this phenomenon can appear at local regions. One of the major reasons is that as in China, top-down management principle guides the general administrative sections, where local governmental constructions always have the tendency to imitate the settings from central and higher hierarchies, in this way, the principle facilitates the task breakdown, claims vertical leaderships and instructions. Evidences have also been found during the interview analysis, for example, after the establishment of KDAB at city level, the respective lower tier county governments also in succession established their own Dianchi Administrative Officers in orders to better coordinate with city tier. Of course, the internal constructions of lower tiers are not identical to higher tiers governments, mild variations could exist. Thus, when looking back at the internal constructions of Kunming City government, in comparison with the internal settings of provincial and county tier governments, city government constructions are slightly more specific, while more concrete tasks and management processes are assigned at city level, and it results in more detailed divisions of related bureaus. In general, when we see Kunming City government as whole, it is comprised of numerous independent and interlinked bureaus, and ahead of those bureaus, the central office also exists, which generally coordinates the working tasks of them.

Beneath city tier, there are three lower administrative levels existing in Kunming City, which are counties, townships and villages, with significant decreasing degree of involvement, each level's administration participates water resource management in its own practices. At these lower tiers, types and formations of their governments vary parallel, these differences are generally due to the different formations of communities, density of population, lifestyles and economic strengths. Regarding county tier, three different kinds of categories could be identified, which are 7 districts (more urbanized with stronger economic strength), 7 counties (less urbanized with lower economic strength) and 3 special districts (recently established for boosting development paces by high technologies, tourisms and investment purposes), in comparison, their governmental internal structures could vary at certain limited degrees, and in general less sophisticated than city level. From the interview analysis, we found that with respect to water resource management, county level is involved in some fundamental water administrations but with certain limitations. On one hand, county tier is engaged in multiple water management tasks, such as monitoring water qualities within their task ranges; conducting water legal controls; implementing central or higher-level policies (such as Five-Year Plan, Major Water Program, and Water Ten Items [4]) assigned projects, And providing services for water supply and wastewater treatment, if required. On the other hand, due to the top-down administrative principle, Kunming City government claims its leadership against county governments, thus, decisions and arrangements could be made that some fundamental water management rights are held rather in hands of city government, for example, counties could monitor quality of water bodies, only when city bureaus would not wish to perform or have assigned to them. Further evidences were found during the interview analysis, such as county tier does not have the rights to issue permissions for water extraction; and county governments could not provide water supply and wastewater treatment services if city government claims the responsibility already. In summary, the engagement of county governments in water sector could be stated that they are actively involved within wide aspects of water management tasks under the leadership and instructions from Kunming city government. Regarding the township and village levels, their degrees of involvement in water management are significantly lower than county level, and in most circumstances, they are responsible for ensuring the functioning of water conservancy facilities and fulfilling assigned duties from county tier.

As local governments are executing concrete administrative tasks, the local party committees besides are playing central roles in leading and coordinating their general performances. Party committees are extensively established at each administrative level and also govern strictly by the top-down principle, they form their own internal structures and the personnel are interwoven with their corresponding government settings. The analysis of the interviews found that at city level, multiple special organizations are formed together by Kunming City party committee and government, such as Dianchi Lake Protection Commission, Kunming Water and Soil Conservation Commission, and Dianchi Integrated Treatment Leading Group. Those organizations are normally led by the top leaders of party committees and government, on one hand they function to monitor the general performance of various management bureaus at same or lower levels, and on the other hand they create the platform to set up strategies, requirements and enhance communications among various water related bureaus and departments.

Abundant Water Relevant Subordinate Authorities Scatteredly Exist

Among general Chinese administrative systems, subordinate authorities exist as very important governmental organizations that are widely established at each level of government. Regarding water sector, those subordinate authorities are normally responsible for conducting concrete technical and professional water tasks on behalf of governmental bureaus, which are normally established attached to them. As with Kunming City, water related subordinate authorities are also established among different city bureaus, while the three most water relevant bureaus, namely KEPB,

KDAB and KWRB, manage many important ones. In this section, we will focus on discussing the interviewed subordinate authorities that are in the field of water research, development and monitoring, and we will use these examples to demonstrate how those water related subordinate authorities function and coordinate with each other.

Interview with Kunming Academy of Environmental Sciences (KAES) tells that firstly, there are two levels of Academy of Environmental Sciences (AES) existing in Yunnan Province, which is the single Yunnan Provincial Academy of Environmental Sciences (YAES), and multiple city level AESs established nearly for each city within Yunnan Province. For YAES, it is a subordinate authority of Yunnan Provincial Environmental Protection Department, while KAES is attached to KEPB, and all other city level AESs are also managed by their corresponding city Environmental Protection Bureaus (EPB). As with KAES, its intrinsic functions are conducting water related research tasks, such as investigating ecological restorations; pollution carrying capacity of specific water bodies; and modeling hydrological situations. On the other hand, due to the governmental characteristic, its primary responsibilities also include providing professional services and consultations to city government for facilitating management purposes, for example initiating and revising different kinds of planning. Thus, comments have been made by the interviewees that KAES is more functioning as an applied research institute, where limited fundamental research work takes place. Another interesting characteristic of KAES indicates that in comparison with other AESs from parallel cities of Yunnan Province, KAES is much well developed and organized than others, which is mainly due to the importance of Dianchi Lake, and in result, KAES has received much more attention and duties regarding Dianchi Lake treatment issues.

The interview with another important city level subordinate authority - Kunming Environmental Monitoring Center (KEMC) revels that firstly, Environmental Monitoring Centers (EMC) are established at three levels, namely province, city and county, and their primary responsibilities are conducting water quality related routine and specific monitoring tasks. For all the different EMCs, they are always established as subordinate authorities under their corresponding Environmental Protection Departments/Bureaus. For KEMC, it naturally conducts and coordinates the general water quality monitoring issues in Kunming City, and in specific, it executes routine water quality monitoring tasks for around 170 measuring points within city territory, where they perform the standard methods that are required by national standards and requirements. With professional equipment and facilities, KEMC could monthly produce monitored water environmental data, nevertheless, it is not able to fully monitor all the 170 required monitoring points, thus the numerous county level EMCs should act and help to fulfill the monitoring tasks if it is requested. In the end, county EMCs need transfer their monitored data to KEMC for comprehensive processing. With all the water quality monitored results, KEMC monthly reports and transfers the necessary data to Yunnan Provincial EMC for further producing reports to national government and general publics. Thus, here we could find that EMCs at different tiers have their own connections and cooperative mechanism, in fact, EMCs at higher level provide consultations, point out mistakes or areas for improvement, request and delegate tasks to lower EMCs, nevertheless, they do not hold the direct leadership to the lower ones, while this power belongs to the corresponding management bureaus. Some further phenomena have been discovered from the interviews at KEMC, for example, interviewees expressed that limited financial and personnel resources have caused great burdens to KEMC's over-saturated routine tasks; automatic monitoring points exist with different management patterns; water biological monitoring has not been maturely developed and only limited parameters are currently been monitored; online released water environmental data is released in very simple formats; and water quantity related monitoring tasks are performed by other independent subordinate authorities.

In summary, across governmental tiers, subordinate authorities widely exist and perform their required tasks, cooperative mechanisms exist among them for accomplishing miscellaneous and concrete water management tasks.

Water Administrative Functions Spread Among Governance Constructions

Water administrative functions are understood as the water relevant official authorities that hold the rights of explaining water related laws, policies and regulations and making decisions as well as processing water management issues, such as authorizing permissions, making punishment decisions, and rejecting water using applications. Water administrative functions could be thus regarded as important contribution for implementing water rights. In order to execute those administrative functions, they should be designed and arranged within their corresponding water governance constructions, and how those functions are performed is crucial for their effective implementations. Our interview study has found some essential characteristics of water administrative functions in Kunming City and will be introduced in this section.

In general, multiple interviewees from KEPB, KWRB and KDAB have emphasized that different administrative functions are spread among different water related bureaus across governmental tiers. Representative examples could be illustrated here that in the first place, there are water law and regulation related departments established in each water related bureau from city and county tier government, and in principle, they have the rights to design some essential local water regulations and policies at the initial stage, then they should be further revised by higher authorities, thus, those departments could work individually as well as cooperating to initiate various policies and regulations, depending on the concrete tasks. The second example is about issuing various water related permissions, which are regarded as essential approaches to reasonably and sustainably regulate anthropogenic activities that could cause adverse effect on water environment. three kinds of water related permissions are regarded significant in Kunming City, which are wastewater discharge permissions, waste discharge permissions and water drawing permissions. For wastewater discharge permissions, they are designed to verify whether the wastewater discharge activities are in accordance with the requirements to connect with existing drainage facilities. The interview found that for 5 of the urbanized counties, the permissions are together assessed and issued by KDAB, nevertheless, for other districts and counties, these permissions are then not granted systematically, while some counties issue them by other bureaus within their own governmental constructions, and some counties even do not issues them but just combine with other different permissions, for example city planning related permissions. Regarding waste discharge permissions, they indicate how much amount of different pollutants, including wastewater are granted to be discharged by specific entities. Depending on the types of waste dischargers, the permission issuing tasks are divided among city and county tier EPBs. In addition, the water drawing permissions are designed to ensure the reasonable water extraction and utilization. Regarding these permissions, county governments have no rights to issues them, while this responsibility is shared by the KWRB and provincial WRB, depending on the dimensions of the water extraction activities. The third example regarding scattered water administrative functions is about executing water legal controls, which are concerned as imperative ways to fight against illegal activities that bring harms to general water environment. In Kunming City, the task forces for water legal controls are distributed among various management bureaus, mainly include KEPB, KDAB, KWRB and their corresponding lower county tier bureaus. At city level, each task force from the 3 mentioned bureaus has its own focused task ranges to conduct water legal controls, for KEPB, it is more responsible for water pollution related incidents, which happen in urbanized regions, as with KWRB, its task force aims at fighting against activities that cause harm to water resource distribution, consumption and allocations, regarding KDAB, its task force has rights to handle comprehensive water legal issues within Dianchi Lake Basin. In fact, besides their independent working patterns, those task forces have their own interlinked mechanism that at necessity, they need to cooperate for investigating and processing more complicated water legal cases. As with county tiers, they hold their major responsibilities to process legal issues independently when the water cases are less significant and more rurally arisen, in addition, they also need to highly cooperate with task forces from city tier to provide local consultation and necessary assistances. Thus, water legal cases are normally resolved by single bureau or joint task forces from water bureaus across management tiers.

Besides the above more in detail mentioned water administrative functions, further examples have been found from the interview analysis and express that those functions are widely performed by joint administrative efforts: KWRB and KEPB both conduct studies on pollution carrying capacities of water bodies for guiding administrative decisions on pollution controls; KWRB and Kunming Land Resources Bureau both manage groundwater by different categories; Kunming Industry and Information Technology Bureau and KWRB both have rights to manage small scaled hydropower plants; various bureaus at different governmental tiers monitor and supervise the wastewater treatment plants (WWTP) in Kunming City urbanized regions; and with the example which has been mentioned in the above section that water monitoring tasks are executed and divided by numerous related subordinate authorities at various governmental tiers.

Highly Integrated Core Water Service Pattern for Kunming Urbanized Regions

In Kunming City, the two water core service tasks, namely water supply and wastewater treatment are provided differently between urbanized and rural regions. In urbanized regions, where five economically better developed counties sit, their core water service tasks are centrally managed and served by large scale city controlled and owned companies. More concretely, at current situation Kunming Dianchi Water Treatment Co. Ltd (KDWT) manages 21 WWTPs that are distributed within Kunming City urbanized regions, and their treatment capacity can satisfy 90% population that settle there; on the other hand, Kunming Water Supply Group Co. Ltd (KWSG) integrally provides water supply service by its 13 subordinate Water Supply Enterprises (WSE), and its supply capacity has exceeded the overall demand from the urbanized regions. Since the water service providing companies have very strong political and governmental characteristics, they fully operate under the leadership of city government and city party committee in means of receiving supports, supervisions and instructions. In fact, special commission, Kunming State-owned Assets Supervision and Administration Commission (KSASAC), has been long established to monitor the governmental controlled and managed companies, it aims at affecting them by appointing important personnel, modifying corporate structures, planning development directions, etc. Thus, in brief, core water services of Kunming City urbanized regions are highly controlled by city government.

As with rural regions, although this study has not made interviews at counties and lower levels in Kunming City, some interviewees have mentioned about the general conditions of their water core services, which could be summarized here after interview analysis. Firstly, core water services are not consistently managed and planned in rural regions, the management patterns of WSEs and WWTPs could vary, such as managed directly by local governmental bureaus, and private or semi-private firms. Secondly, the scales and technology competence of water service firms are smaller and less advanced in rural regions, especially in rural townships and villages, it could happen that only very basic facilities are installed for water services purposes.

Existing Water Management Processes Have Potentials for Further Refinement

As has been introduced in Sect. 1.2.3.1.3, four representative water management processes have been analyzed, although with these few study examples, some very phenomenon process related characteristics could be identified and will be expressed in this section.

We have found that all these four studied management processes did not have previous ready-made comprehensive flow charts, thus it has required us to talk to different involved officers to clarify stepwise each single detail of studied water management process. After analyzing these collected process data, we firstly identified that fractional administrative functions of various bureaus could affect the sequential logic of different water management process, for example, within the process of WWTPs monitoring, continuous process has been split and interrupted due to diverse management scopes of different authorities, thus multiple indepen-dent small processes together can just tell the full mechanism of WWTPs monitoring. Secondly, it has been recognized that flaws within one specific water management process could be complemented by other similar or relevant process(es), the example goes that when we parallel investigated through the wastewater discharge and waste discharge permissions issuing processes, not one single process could entirely tackle the issue of granting wastewater discharge permissions, nevertheless, some flaws from both of the two related processes could be considerably complemented by each other. In addition, it has also been noted that within each studied management process, one or more executive steps of them were not thoughtfully designed or elaborated, while those steps were either not well defined by its description of step content or their logical sequences have not been clearly determined. Finally, from the example of ecological compensation implementation process, we found that when with primary governmental attentions, its action steps tend to be well designed in terms of logic and depth of details.

With those identified water management process characteristics, we could at least view that those required tasks have been executed in varied degrees of consum-mations, in the meanwhile, multiple potentials for further refinement have been recognized through their analysis.

Dianchi Lake Witnessed the Continuous Evolvement of Governmental Constructions

Referring to water resource management in Kunming City, the Dianchi Lake (Basin) has always played a very significant role in affecting the city's actions and policies for governing its water resource in general. Along history, Dianchi Lake has been important for the economy, tourism development as well as many aspects of residents' livelihood of Kunming City, and its severe pollution problem has attracted enormous governmental attentions and induced abundant treatment actions since the beginning of pollution (around 1986) till today. In fact, Dianchi Lake has witnessed some essential evolvement of governmental constructions, which will be introduced in this section.

Firstly, in 2002, KDAB was established in order to integrally resolve the severe pollution problem of Dianchi Lake, at that time, it was a very vanguard act to create a city bureau for managing a lake within the range of a city territory. Till today, KDAB is one of the most important core governmental bureaus of Kunming City, which plays significant roles in many aspects, including constantly coordinating with other traditional established bureaus. Secondly, for the better pollution treatment of Dianchi Lake, Dianchi Lake Protection Commission and the Dianchi Integrated Treatment Leading Group have been successively established, both are made up of city governmental and party committee high level leaders. With these two special commissions, the speed and coordinative efficiencies for Dianchi Lake pollution treatment has been greatly advanced. Thirdly, with the massive national policy River Master Program [7] been widely implemented all over China, Kunming City also created a special office named Dianchi Lake Basin River Master Program Office, which aims at thoroughly implementing River Master Program for Dianchi Lake

Basin. In addition, as also mentioned above that due to the continuous treatment demands for Dianchi Lake, KAES has received amount of treatment and remediation related tasks along the historical stages, till today, the importance of Dianchi Lake has greatly influenced the internal constructions of KAES that more employees have been recruited, the department dealing with lake studies has internally become the most important one, etc.

Water Governance Structural Characteristics of Bavaria
Systematically Designed Water Governance Constructions

As has been shown in Sect. 1.2.3.1.3 that the water governance system of Bavaria generally consists of 4 governmental levels and 2 types of technical oriented organizations, namely LfU and WWAs (Figs. 1.1, 1.2, 1.3, 1.4, 1.5, 1.6, 1.7, 1.8, 1.9), and their essential functions have been also briefly stated above. In this section, further fundamental governance structural characteristics will be demonstrated after interview analysis.

Beginning with StMUV, as the highest state authority for water resource management in Bavaria, it governs and influences water administration at strategic and diplomatic level. Internally, StMUV is comprised of multiple departments, and the Department 5, namely Department of Water Resource Management and Soil Protection is established with 10 working units that tackle water governance from different

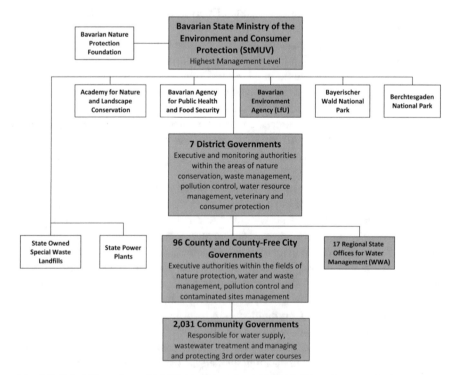

Fig. 1.9 Brief illustration of highlighted Bavarian water administrative organizations among general environmental protection governance system (modified by authors, StMUV n.d.)

concrete water aspects, such as water laws, national and international river basins, monitoring issues, water supply, and wastewater treatment. With this department formation, the StMUV is responsible for several major concrete water governance tasks. Firstly, it interprets and implements European and federal water legislations at state level, in the meanwhile, it also initiates and releases state water laws, regulations, policies and makes important administrative decisions for whole Bavaria. Thus, in principle, StMUV conceptually forms central strategies regarding water management in better and more holistic concepts. Secondly, since important water bodies (1st and 2nd order water bodies (LfU n.d.d)) are managed at state level, StMUV is then directly in charge of their development strategies, including generally regulating and supervising hydraulic constructions as well as initiating water body protection plans. In addition, StMUV oversees the management conditions of water supply, wastewater treatment within whole state and provides financial support for those enterprises on basis of necessity. Furthermore, StMUV actively involves within various national and international river basin commissions and maintains steady relationships with external and foreign bodies regarding sharing water resource management duties, responsibilities and rights.

Beneath ministry level, 7 district governments act as upper public water administrative authorities in Bavaria. Internally, each district government is comprised of multiple working branches, and the water administrative sector at district level is, for each district, identically established as working unit "Water Resource Management" under the working branch "Environment, Health and Consumer Protection". In general, water administration at district level is responsible for executing, authorizing, coordinating and monitoring various water issues that connect higher and lower tiers. More specifically, district water authorities at first directly supervise and monitor the activities of WWAs that are situated in the same district territories, they have the rights to apply funding from ministries for their WWAs and pre-examine important projects and plans initiated from them. Secondly, district water authorities monitor county governments at lower tiers for ensuring their water rights have been appropriately performed. Thirdly, they also have rights to fulfill administrative functions, when concerning more significant and cross-regional issues, at these conditions, district governments play roles in providing coordination and finding balanced resolutions. In addition, district water authorities should execute various orders and requests from higher ministries.

Under district level, 71 county and 25 county-free city governments form administrative nets for water governance at county tier. Since each county enjoys its power of self-administration, it has freedom to design its own suited governmental constructions, and results in internal department settings in various styles. Nevertheless, water rights unit is always built within each county government (normally under environmental related department) for fulfilling its responsibility of processing water related laws, statutes and legal matters, more details have been already introduced in the Sect. 1.2.3.1.3.

Ranked as the smallest administrative divisions, 2,031 community governments are formed throughout Bavaria. Since communities enjoy much entitled sovereign rights and administrate more independently than counties, their governmental constructions also differ according to local management conditions. Usually, different types of unions and firms are locally formed for providing communities water supply and wastewater treatment services, and practical working units are normally formed within community governments for the purpose of managing and protecting 3rd order water courses (LfU n.d.d).

By knowing in depth about the water governance construction in Bavaria and in combination with interview analysis, two prominent structural characteristics have been found. In the first place, although water management authorities in Bavaria are geographically distributed among several governmental tiers, they are usually regarded collectively as within one water governance system, where their hierarchies and functions have been officially illustrated and explained (StMUV n.d., Fig. 1.9). On the other hand, Human Resource Management for water administration in Bavaria is regarded as one independent system with its own rules on personnel mobility, it keeps on concerning personnel statistics and executing management measures for bringing personnel through diverse positions within Bavarian water administrative system. The second discovered water governance structure feature tells that the water governance style is uniform throughout Bavaria, it could be reveled from the facts that water rights units are widely and equally established for parallel local governments, although their respective governmental settings could vary to certain degrees, in addition, standard management measures and processes have been created, which are identically implemented at local conditions. Moreover, WWAs are distributed in the entire state to ensure each local region could receive same quality of water services.

Formation of Integrated Technical and Scientific Oriented Water Institutions
In Sect. 1.2.3.1.3, the functions of WWAs and LfU, namely Bavarian technical and scientific oriented water institutions, have been briefly stated. In this section, their organizational and more management patterns will be further illustrated.

Starting with WWAs, they exist with distinctive Bavarian characteristics, while most of other states in Germany do not establish statewide official institutions that only focus on conducting water management technical tasks. In total, 17 WWAs are evenly distributed in Bavaria, which guarantees that each county or county-free city could receive their equal consultative water management services, in addition, the distribution of WWAs also ensure fundamental water tasks could be parallel and locally executed. All WWAs are similarly formed in matrix styles, this special type of internal constructions enables that on one hand, each county or county-free city could have its own contact persons in its corresponding WWA, since the departments of WWAs are horizontally installed for providing service to one more multiple counties. Under those departments, working units are formed to conduct different categories of water tasks, namely water quality, quantity related and biological monitoring issues; water supply, groundwater and soil protection aspects; wastewater treatment as well as hydraulic constructions and water body development. On the other hand, based on those categorized tasks, vertical technical branches are set up, which constantly provide internal professional consultations to those working units at necessity. Besides resolving technical water tasks, WWAs are also in close connection and communication with diverse relevant social and public entities, such as Isar-Allianz (Isar-Allianz n.d.), Bund für Umwelt und Naturschutz Deutschland (BUND n.d.), state and local fishery associations, etc. WWAs on one hand receive complains, ideas and professional advices from those public entities, and on the other hand, WWAs also inform them in time about the administrative decisions, which are significant to their interest. In addition, WWAs are in first line having communication with private citizens, when citizen address their complains, doubts and questions to them and they should accordingly provide feedbacks to those citizens in a timely manner.

On the other side, LfU is established as subordinate authority to StMUV and stands as the integrated and highest water scientific institute for whole Bavaria, and being as the only official environmental agency, no further similar ones are formed at lower level governments. Internally, LfU consists of 10 departments tackling all environmental aspects, and among them, 5 departments are fully or partially engaged in numerous concrete water issues, ranging from water services, laboratory analysis and evaluations to research and development. Functioning as knowledge hub for water resource management in Bavaria, LfU firstly provides scientific consultations to StMUV prior to develop water development goals and strategies; secondly, LfU ensures the basis of WWAs' technical performances by supporting them with guidelines, practical measures and instructions when confronting technical difficulties; in addition, LfU is also available to water rights units in county governments for assisting to make proper administrative decisions; moreover, LfU centrally obtains and processes water data and further provides to publics for their own utilizing purposes.

With in-depth knowing about Bavarian water technical and scientific institutions, their establishment indicates the characteristic that without complicated institutional arrangement, water scientific and technical tasks are centrally executed by two types of institutions, namely WWAs and LfU, they broadly communicate with each other, local governments at different tiers as well as social entities to resolve practical water management issues.

Decentralized Core Water Service System

In Bavaria, there are currently around 2,300 WSEs, 3,250 water protection zones, 9,000 groundwater abstraction sites, 2,640 public WWTPs and 100,000 small scaled WWTPs distributed throughout the entire state territory [40]. Endowed with high power of self-administration, municipalities in Bavaria enjoy the sovereign rights of managing their own water supply and wastewater treatment services. In principle, it has been kept that core water services in Bavaria are decentrally and locally provided, since multiple advantages are believed to be gained by adopting this type of management pattern. In the first place, WSEs could be formed to adapt local environmental and socio-economic conditions, thus those enterprises, usually organized at small to medium sizes, could be more flexible to optimize themselves for fulfilling local requirements. Secondly, with local services, water is supplied with short distances thus could be delivered fresh to consumers, on the other hand, wastewater could also be swiftly transferred to WWTPs. In addition, source water could be better protected and easier to be traced, regarding WWTPs, risks could be reduced when confronting incidents.

Depending on local conditions, namely types and sizes of municipalities, population as well as industrial and residential settlements, water enterprises are organized in different ways. For larger cities, public firms are often established to handle water service tasks, those firms could be fully operated by local governments or managed with mixed type of business modes; in other conditions, those tasks could be undertaken by private companies, which should operate under the supervision of their corresponding governments. Regarding smaller sized communities, special purposed associations are usually formed to provide water services for single or multiple adjacent communities, where cooperation is recognized at necessity among them.

In order to enhance the operation performance of those abundant and scattered water service enterprises, special "neighborhood" program was officially launched [40], which provides professional communicating and training platforms for water enterprise employees from neighbor communities. Under this program, routine meetings are held to covey technical knowledge, resolve common difficulties and further train their personnel. Thus, performance qualities of the overall water enterprises could be considerably maintained even without great amount of administrative efforts.

Maturely Designed Water Management Processes

Alike the water management process study for Kunming City, we have also investigated 4 representative processes (see Sect. 1.2.3.1.3) for Bavaria, the analysis has identified several phenomenon characteristics and will be illustrated in this section.

In the first place, we have learned that numerous water management processes, including the 4 studied ones, have been previously designed within Bavarian water governance system, those processes are illustrated in form of 2D flow charts, which not only stepwise demonstrate the process sequences, but also identify each action step's responsible authority. Several interviewees in our study have expressed and believed that those standard designed process flow charts could provide solid guidance during practical implementations, and managers could quickly and clearly monitor the performance quality of those processes at operation. Secondly, the representative studies recognized that the action steps of each management process are usually in detail described, especially for those on-site activities, for example, working aids are concretely and in detail formulated by LfU for those Private Water Management Experts (PSW), who are engaged in officially authorized practical tasks, such as on-site monitoring WWTPs, and conducting laboratory evaluations. Besides, when the management processes are relevant to general publics, the corresponding information is usually released online or in other forms, which could be reached by public stakeholders for their preparation and better engagement, for example, leaflets are produced by LfU for guiding social entities and firms to apply for wastewater discharge permissions. Thirdly, it has been further noted that executive steps within water management processes are usually reflections of relevant laws and regulations.

On one hand, the general goals, namely keeping water bodies in good ecological and chemical status, are often considered and reflected by concerned management processes, for example, when WWAs are evaluating the wastewater discharge applications from wastewater discharges, they heavily consider how those activities could adversely affect the water bodies regarding their current and target natural conditions; further example from the process of handling accidents with water hazardous substances require that the identified polluters should take actions to remedy pollutions for returning the affected water bodies as close to their previous natural statues. One the other hand, there are specific laws and regulations already existing, such as Bavarian Administrative Procedures Act (BayVwVfG), and Administrative Procedure for the Implementation of Water Law (VVWas), which specify the designed water management processes should adequately reflect their requirements. Lastly, our study found that continuous discussions and debates are held to optimize the outdated or controversial executive steps, for example, there are discussions regarding whether it is appropriate to inform the WWTPs' operators shortly prior to their on-site monitoring practice.

Challenges of Bavarian Water Governance Requires Long Term Efforts

In this study, besides the effective water management measures were broadly identified, we have also learned multiple prominent challenges that Bavarian water governance system is confronting.

Although much effort has been devoted to advancing and optimizing its water governance construction, several aspects, such as non-point source pollutions, groundwater contamination and treatment, micropollutants and new water hazardous substances in water bodies, reclamation of energy and materials, and climate change induced issues are continuously bringing challenges to Bavarian water management system. Since those aspects could not be resolved at once and always evolve at different historical stages, official authorities thus should always invest to develop and equip refined professional knowledge and technical measures to bring better answers and solutions.

Challenges also exist when there is necessity to negotiate with private entities for advancing water protection and development measures. Very typical example was given by the discussion with WWAs that in order to further implement natural restoration plans, considerable lands should be obtained and purchased from private owners, during this process, general doubts, objections as well as increasing prices of real estate are very typical obstacles that slow down those management plans.

In Bavaria, there are long-lasting conflicts existing between water and agriculture sections. One on hand, increasing and intensified agricultural activities constantly bring considerable amount of contaminations to groundwater and surface water, although efforts from water administrative authorities were devoted to intervening, limited success was achieved. The reason lies in that agriculture is regarded as crucial industry in Bavaria, their social unions together with administrative authorities thus possess very strong political voices, various special regulations were formulated in favor of their economic growth. The Bavarian water authorities consider that the breakthrough of effectively controlling pollution from agriculture sector not only requires stronger water administrative interference, but it also counts on the self-awareness from agriculture governance sections. On the other hand, in order to maintain reliable biological monitoring performance in Bavaria, fish data is very essential to the reach of water sectors, nevertheless, significant fish data is primarily held in hands of agriculture authorities, additional efforts are normally required to obtain those data by water institutions.

Furthermore, the recent personnel reform at Bavarian administrative system has greatly reduced the number of civil servants at water sector, which resulted in difficulties to reassign tasks on limited personnel for maintaining reliable water governance performances.

Comparison of Water Governance Structural Characteristics Between Kunming City and Bavaria

Essential water governance structural characteristics of Kunming City had Bavaria have been in detail illustrated in sections above. In this section, their fundamental features will be further summarized and compared in Table 1.7. Under different political and administrative systems, Kunming City and Bavaria also govern water resources differently in terms of management philosophies as well as concrete management measures. Nevertheless, it could not be claimed that any local water governance system is fully well established, which does not require further optimizations. Therefore, from our study, those comparisons between Kunming City and Bavaria could be used to bring inspirations for their further respective water governance optimizations.

Table 1.7 Essential water governance structural characteristics comparison between Kunming City and Bavaria

Comparison keywords	Kunming city	Bavaria
General governance principle	Top-down governance principle prevails, which indicates central and higher-level governments generally lead administrative affairs of lower ones	The subsidiarity principle prevails, which indicates governance tasks should be distributed with all their rights and obligations at the lowest possible efficient level, thus it ensures local self-administration power
Internal structures of local parallel governments	Internal structures of local parallel governments are constructed with great similarity but could differ at varying degrees due to defined types and economic strengths	Internal structures of local parallel governments vary with larger differences (besides district level governments) due to self-administration rights

(continued)

Table 1.7 (continued)

Comparison keywords	Kunming city	Bavaria
Alikeness of cross-tier governmental constructions	Lower tier governments significantly imitate the internal structures of higher tiers	Local governments design their own internal structures, no specific cross-tier imitation patterns have been found
Water authority at each local government	Multiple water related bureaus exist at each local government and jointly administrate miscellaneous water issues	Only one water department/rights unit is formed at each local government for administrating water issues
Distribution of water administrative functions	Water administrative functions are distributed among numerous water related governmental departments/bureaus across hierarchies	Water administrative functions are separately held in hands of water rights units from corresponding local governments
Political influence on water administration	Party committees set strategies for local water development and play central roles in leading and coordinating significant water governance issues	Bavarian water governance system appreciates professional and technical administrative foundation, which accepts limited and necessary influence from political field
Water technical and scientific authorities	Numerous water technical and scientific authorities are organized as subordinated authorities attached to various governmental bureaus across administrative tiers, which independently or jointly execute technical and scientific water tasks	Two types of water technical and scientific institutions (WWAs and LfU) support and provide services and consultations to cross-tier local governments, and together execute integral technical and scientific water tasks
Core water service patterns	Water resource is centrally supplied, treated in urbanized regions and less systematically organized in rural regions	Water resource is in principle locally, namely within the reach of respective municipalities, utilized, treated and protected
Water service enterprises	City owned state companies centrally provide water service for urbanized regions, while governmental controlled enterprises at lower tiers provide services for rural regions, which could be organized in different business forms	Ensured with power to locally provide their own water services, municipalities have established abundant self-controlled small to middle sized water service enterprises

(continued)

Table 1.7 (continued)

Comparison keywords	Kunming city	Bavaria
Pre-designed water management processes	Significant water management processes are less systematically designed in various forms of charts	Significant water management processes are systematically pre-designed in flow charts, which in detail indicate action steps and each step's responsible authority
Correlative water management processes	Correlative water management processes generally exist, which complement each other and could also result in task overlapping	Water management processes are usually designed more comprehensively for resolving independent water tasks
Public participation	Public participation is adequately recognized, while some water related information is released to public in less concrete forms	Public participation is well recognized and treated by Bavarian water administration

1.1.3   Conclusion

In this study, qualitative interviews were conducted among water governance internal systems of Kunming City and Bavaria, those interviews were aimed at in-depth learning about their respective water governance constructions, concrete management duties, performance strengths as well as confronting challenges. By adopting grounded theory analytical methodology, their most essential water governance characteristics were in detail demonstrated and discussed, in addition, fundamental water governance features of Kunming City and Bavaria were comprehensively compared.

References

1. Ahmady GA, Mehrpour M, Nikooravesh A (2016) Organizational Structure. Procedia Soc Behav Sci 230:455–462
2. Bayerisches Landesamt für Statistik (2017) Bevölkerungsstand. https://www.statistik.bayern. de/statistik/bevoelkerungsstand/. Accessed 27 July 2018
3. BUND (n.d.) Wir über uns. https://www.bund.net/ueber-uns/. Accessed 24 June 2019
4. China Water Risk (2015) New 'Water Ten Plan' to Safeguard China's Waters. http://www. chinawaterrisk.org/notices/new-water-ten-plan-to-safeguard-chinas-waters/. Accessed 24 June 2019
5. China Water Risk (2016) China's 13th Five-Year Plan for Ecological & Environmental Protection (2016–2020). http://www.chinawaterrisk.org/notices/chinas-13th-five-year-plan-2016-2020/. Accessed 24 June 2019
6. Correia FN (2003) Institutional water issues in Europe. World Water Congress 11
7. Dai L (2015) A new perspective on water governance in China: Captain of the River. Water International 40:87–99

8. Das Gupta A, Singh Babel M, Albert X, Mark O (2005) Water sector of bangladesh in the context of integrated water resources management: a review. Int J Water Resour Dev 21:385–398
9. Dohmann M, et al (2016) German contributions to the Major Water Program in China: "Innovation Cluster–Major Water". Environmental Earth Sciences 75
10. Evans JW, et al (2006) Environment matters at the World Bank: 2006 annual review (English), Washington, DC
11. Grambow M (2008) Wassermanagement: Integriertes Wasser-Ressourcenmanagement von der Theorie zur Umsetzung, 1st edn. Friedr. Vieweg & Sohn Verlag, Wiesbaden
12. Grambow M (ed) (2013) Nachhaltige Wasserbewirtschaftung: Konzept und Umsetzung eines vernünftigen Umgangs mit dem Gemeingut Wasser. Vieweg+Teubner Verlag, Wiesbaden
13. Grambow M, Disse M, Chen K, Patalong H, Uhl H-D (2019) Sustainable Water Resource Management in China – Reflections from a Comparative Governance Perspective. In: Köster S, Reese M, Zuo J'e (eds) Urban Water Management for Future Cities, vol 12. Springer International Publishing, Cham, pp 283–301. doi: https://doi.org/10.1007/978-3-030-01488-9_13
14. Hassenforder E, Barone S (2018) Institutional arrangements for water governance. Int J Water Resour Dev 21:1–25
15. He J, Xu X, Yang Y, Wu X, Wang L, Li S, Zhou H (2015) Problems and effects of comprehensive management of water environment in Lake Dianchi. J Lake Sci 27(2):195–199
16. Isar-Allianz (n.d.) Grundsatzpapier, Ziele und Organisation der Isar-Allianz. http://www.isar-allianz.de/aufgaben.htm. Accessed 24 June 2019
17. Jin X, Wang L, He L (2006) Lake Dianchi. Experience and Lessons Learned Brief:159–178
18. Kilgariff E, Fernando R (2005) The GeForce 6 series GPU architecture. In: Fujii J (ed) ACM SIGGRAPH 2005 Courses on - SIGGRAPH '05. ACM Press, New York, USA, p 29
19. Koohborfardhaghighi S, Altmann J (2017) How organizational structure affects organizational learning. J Integr Des Process Sci 21:43–60
20. Kunming City Government (n.d.) Kunming City Government Official Internet Portal. http://www.km.gov.cn/. Accessed 24 June 2019
21. LfU (n.d.d) Allgemeine Daten zur Wasserwirtschaft. https://www.lfu.bayern.de/wasser/allgemeine_daten_wasserwirtschaft/index.htm. Accessed 24 June 2019
22. LfU (n.d.a) Amtliche Festsetzung von Überschwemmungsgebieten. https://www.lfu.bayern.de/wasser/hw_ue_gebiete/amtliche_festsetzung/index.htm. Accessed 27 July 2018
23. LfU (n.d.b) Das weiß-blaue Klima. https://www.lfu.bayern.de/wasser/klima_wandel/bayern/index.htm. Accessed 27 July 2018
24. LfU (n.d.c) Mittelwerte des Gebietsniederschlags. https://www.lfu.bayern.de/wasser/klima_wandel/bayern/niederschlag/index.htm. Accessed 27 July 2018
25. LfU (2010) Wasserschutzgebiete für die öffentliche Wasserversorgung
26. LfU (2017) Flüsse, Seen und Grundwasser in Bayern - Gewässer auf dem Weg zum guten Zustand
27. Liu J, Li J, Li L, Wang Z, Chen S (2017) Comprehensive analysis on spatial pattern of water resources shortage in Kunming City. J Yangtze River Sci Res Inst 34:6–10
28. Liu R, Cao X (2017) Pollution treatment of practices of Dianchi Lake in the past twenty years. Environ Sci Surv 36:31–37
29. Meng T (2016) Application of integrated river basin management in Dianchi basin. Chin J Environ Manage 3:53–59
30. Merriam SB (2009) Qualitative research: A guide to design and implementation / Sharan B. Merriam, Rev. ed. Jossey-Bass, San Francisco, Calif.
31. Ministry for the Environment of New Zealand (2009) How Does the National Environmental Standard Apply to Water and Discharge Consents? http://www.mfe.govt.nz/publications/rma/draft-users-guide-national-environmental-standard-sources-human-drinking-water/4. Accessed 27 July 2018
32. Ministry of Ecology and Environment of China (n.d.) Major Science and Technology for Water Pollution Control and Treatment. http://nwpcp.mep.gov.cn/. Accessed 24 June 2019

33. Ministry of Water and Environment of Uganda (n.d.) Institutional Framework for Water and Environment. http://www.mwe.go.ug/mwe/institutional-framework-water-and-env ironment. Accessed 27 July 2018
34. OECD (2014) Water Governance in the Netherlands. OECD Publishing
35. OECD (2015) OECD Principles on Water Governance
36. Pahl-Wostl C, Conca K, Kramer A, Maestu J, Schmidt F (2013) Missing Links in Global Water Governance: A Processes-Oriented Analysis. Ecology and Society 18
37. Saleth RM, Dinar A (2013) The Institutional Economics of Water: A Cross-Country Analysis of Institutions and Performance. World Bank and Cheltenham, UK: Edward Elgar, Washington, DC
38. Sengupta PK (2018) Industrial water resource management: Challenges and opportunities for corporate water stewardship. Challenges in Water Management. John Wiley & Sons, Ltd, Chichester, West Sussex, Hoboken, NJ
39. StMUV (n.d.) Organisation im Bereich Wasserwirtschaft. https://www.stmuv.bayern.de/the men/wasserwirtschaft/wasserwirtschaft_in_bayern/organisation.htm. Accessed 24 June 2019
40. StMUV (2014) Bavaria, Land of Water: Sustainable Water Management in Bavaria. Bayerisches Staatsministerium für Umwelt und Verbraucherschutz, Munich
41. Wang Y (2018) Assessing Water Rights in China. Water Resources Development and Management. Springer Singapore, Singapore
42. Woodhouse P, Muller M (2017) Water governance - an historical perspective on current debates. World Dev 92:225–241
43. Yan F, Daming H, Kinne B (2006) Water resources administration institution in China. Water Policy 8:291–301
44. Yu X, Geng Y, Heck P, Xue B (2015) A review of China's rural water management. Sustainability 7:5773–5792
45. Yunnan Provincial Bureau of Statistics (2015) Yunnan statistical yearbook. China Statistics Press, Beijing

Chapter 2
Ecological Indicators for Surface Water Quality - Methodological Approaches to Fish Community Assessments in China and Germany

Sebastian Beggel, Joachim Pander, and Jürgen Geist

2.1 Introduction

This chapter provides an overview about the implemented water quality assessment regulations and methods with a special focus on Germany and China. In the first part, the general situation and the currently used ecological indicators are presented. In the second part, we exemplarily discuss the fish community assessment by different methodological approaches. Specifically active and passive fishing techniques are compared, which both, single or in combination, are used in Chinese or European fish monitoring surveys.

2.1.1 The Need for Ecosystem Health Assessment

Human activities and consequently the impact on freshwater ecosystems has increased significantly within the past decades, resulting in an unneglectable decline in biodiversity and ecosystem integrity worldwide [13]. As a matter of fact, aquatic habitats associated with 65% of continental discharge have been recently classified as moderately to highly threatened [52, 60, 61], which demonstrates that aquatic ecosystems have become highly dysfunctional due to high anthropogenic pressure. As summarized by Maloney [35], there has been a 81% global decrease in freshwater

S. Beggel (✉) · J. Pander · J. Geist
Aquatic Systems Biology Unit, TUM School of Life Sciences, Technical University of Munich, Mühlenweg 22, 85354 Freising, Germany
e-mail: sebastian.beggel@tum.de

J. Pander
e-mail: joachim.pander@tum.de

J. Geist
e-mail: geist@tum.de

© The Author(s) 2022
M. Dohmann et al. (eds.), *Chinese Water Systems*, Terrestrial Environmental Sciences,
https://doi.org/10.1007/978-3-030-80234-9_2

vertebrate populations over the last 30 years (see also [36]) and an average of 60% of protected species and 77% of habitat types are considered to be in unfavourable conservation status in Europe, with an even higher proportion in rivers, lakes and wetlands [19]. Furthermore, as the human population grows, human contact with and impacts on aquatic ecosystems increase. At present, more than 50% of the human population lives within 3 km of a freshwater system [26]. This has raised concerns over the impacts of anthropogenic activities on aquatic environments [59]. During the past centuries, the requirements for human economic development were often prioritized, sometimes neglecting negative impacts of human activities on aquatic ecosystems and the services they provide. Progress to align both, development and protection of freshwater resources was made during the last decades and the urgent necessity to preserve and restore a good ecological status of aquatic habitats has been recognized [19]. Various monitoring programs to determine water quality status and need for action were developed and implemented worldwide, with the USA and also European countries such as Germany. Recently, countries with a rapid economic development, such as China, are increasingly implementing monitoring tools for ecosystem health status assessment for environmental protection and to preserve biodiversity.

2.1.2   Current Situation in China and Germany

China's rapid economic and social development during the last decades resulted in increasing anthropogenic pressure on freshwater resources. Therefore already in the 11th Five Years Plan (2006–2010) China consistently improved its measures to control pollution and protect the environment within their major watersheds. Major threats to aquatic ecosystems in China are mainly caused by large scale hydromorphological modifications, water scarcity and severe water pollution from industrial, agricultural and municipal emissions. According to the 2017 Report on the State of the Ecology and Environment in China, 23.8% of 1,940 monitored surface water sections were rated with grade IV and V –utilizable for industry, agriculture and recreation without skin contact only– whereas 8.3% did not meet grade V standards –not suitable for human contact or any use- (Table 2.1; [38]). As a developmental goal it was legally implemented in China to improve surface water quality until 2020, with less than 5% classified as grade V or worse, and more than 70% shall meet grade III or better [42].

Historically, European countries such as Germany had a long history of water management and protection. European countries were facing similar environmental problems during their rapid economic development in the 20th century. In the 1970s, when those countries' surface waters were highly polluted, measures were implemented to improve the environmental conditions sustainably, e.g. by improving waste water treatment and regulating emissions into surface waters. A Europe-wide approach was finally implemented in 2000 by the European Union, the EU Water Framework Directive (WFD) [16], where the good ecological status or potential

Table 2.1 Chinese water quality grades and respective water uses (Source: [38])

Grade	Use
I–II	First-class protection zones for drinking water resources; habitats of rare aquatic species; fish and shrimp spawning grounds
III	Second-class protection zones for drinking water resources; fish and shrimp overwintering ground; migration channels; aquaculture areas; swimming sites
IV	Industrial use; recreational purpose (without skin contact)
V	Irrigation (agri-/horticulture, landscape)
V+	Hardly any function except regulating local climate

for surface water bodies was defined as overall goal to be reached by the year 2015. Nevertheless, European countries have largely failed to reach the initial goals in water quality status of surface waters by today. Only 7.7% of Germany's rivers reached the European Water Framework Directive's binding environmental objective of good ecological and chemical status by 2015. 34.4% of the 8,995 monitored sites exhibited a poor ecological status–major deviations from surface water body type-specific values of biological quality elements and biological communities. Further 19.9% were classified as bad–severe alterations to surface water body, type-specific values of biological quality elements and absence or strong degradation of the ecological integer biological communities.

While Germany utilizes biological, hydromorphological and physico-chemical metrics to determine water quality since the early nineties of the last century, China's water monitoring is predominantly based on physico-chemical variables using a dense net of automated monitoring stations covering the main river basins. Those analyses are the most direct approach to define chemical water quality, but on the other hand they are of limited use to allow predictions about aquatic organisms' responses to pollution nor pollution-related influences on ecosystem level [50]. Biological monitoring can provide insights into the organism-habitat relationship by means of indicator species. The presence or absence of such species enables the detection of pollution and the evaluation of biological threats to ecosystem health [23]. Since aquatic organisms cope with chemical, physical and biological habitat alterations over their entire lifecycle, they additionally allow identification of the long-term ecological status of environments. Furthermore, steep pollution and disturbance gradients are appropriately determined by biological data [21]. Acknowledging the relevance of ecological methodologies for water quality assessment, China launched a pilot national monitoring program to assess ecological integrity in 2010.

2.1.3   Ecological Indicators of Ecosystem Integrity

The biological integrity of freshwater ecosystems, here river or lotic systems, is typically considered by separately assessing so-called biological quality elements (QEs) such as fish, macroinvertebrates, macrophytes and algae. In combination these quality elements are used to create a comprehensive picture of the current status of the respective system with the component of the worst assessment determining the status class (one-out-all-out principle). Each of these biological groups can be utilized to derive structural indicators according to their presence or absence in the assessed ecosystem [45]. This allows to draw conclusions about the systems current condition, e.g. good, bad or poor, and additionally facilitates the comparison of the current status with past and future conditions for monitoring purposes. Structural indicators are used to determine ecosystem integrity and allow to draw conclusions about the stability of the ecosystem, e.g. the capacity of an ecosystem to maintain its organization, in rivers mostly linked to river dynamic processes such as sediment relocation or deadwood dynamics. From a scientific point of view, most commonly used indicators are, for example, species richness and diversity, characteristic species assemblage or trophic structure [35]. For monitoring purposes these attributes are especially useful since their assessment can be realized by point-in-time measurements within the respective habitats [43]. Diversity for instance is measured by sampling organisms from a specific community (e.g., fish, aquatic invertebrates, macrophytes) and considering the number of unique taxonomic groups and the number of organisms per group. By applying a standardized sampling design following pre-determined methodological protocols a spatial and temporal comparability is warranted. In river ecosystem assessment, the fish species diversity at a representative site is a structural metric that reflects ecosystem integrity and availability of typical habitats as well as habitat connectivity. Variations in diversity are identified by comparing the current abundance and density of aquatic species in an ecosystem with that of a reference site or historical ecosystem state. Furthermore, this metric can be used to determine potential risks to ecosystem integrity if significant changes in biological structure are detected or specific taxa are missing [31], or on the contrary to measure restoration success [45]. Other commonly used structural metrics include biomass, species richness or dominance or presence or absence of indicator species. Hydromorphological and water physico-chemical criteria are additionally used to aid the biological metrics.

2.1.4 Biological River Assessment Approaches Applied in China and Germany

Legal and institutional frameworks as described in the previous Chapter 1 highlight that management, protocols, analyses and reports of river monitoring results vary between states, commissions, districts and municipalities. Therefore, the comparison of quality elements (QEs) used in river assessments in Germany and China is based on surveys carried out nation-wide. This includes China's Environmental Quality Standards for Surface Water [39] and river monitoring under the WFD in Germany.

2.1.4.1 China

China's rivers are assessed according to the national Environmental Quality Standards for Surface Water [39]. This standard lists 109 chemical and physical parameters of which 24 are classified as basic parameters, five as supplementary parameters, and 80 as specific parameters. Basic parameters are implemented for rivers, lakes, channels, and reservoirs, whereas supplementary and specific parameters are applied to drinking water resources. The standard's basic parameters comprise heavy metals, nutrients, water temperature, pH value, dissolved oxygen, chemical oxygen demand (COD), and biochemical oxygen demand (BOD). For each of the basic parameters, the standard sets thresholds, pursuant to the standard's surface water quality grades (Table 2.1). The grade is attributed on basis of the one-out-all-out principle. Thus, the worst grade attained by any considered parameter represents the final water quality grade. Grade I and II are generally in agreement with the WHO-Drinking Water Guideline values and most recent Chinese Drinking Water Standards [18]. The standard further provides general guidelines pertinent to monitoring procedures and mandate local WRABs to implement and supervise its adaption.

The research process on river health in China was initiated in 1992, when [67] described water quality of Jiuhua River using the Ephemeroptera, Trichoptera, Plecoptera (EPT) and the Family Biotic Index (FBI). Since the early 2000s, researchers in China [62, 63, 70] started to focus on urban river health assessments in relation to China's fast urbanization process. Furthermore, the administrative agencies responsible for the Yellow River, Yangtze River and Pearl River Basin suggested concepts to promote river health [22, 30, 69]. Broader studies on the ecological status and water quality of rivers based on macroinvertebrates were conducted and pilot schemes for monitoring and assessment of those were implemented. These include, amongst others, [28, 44] and [64]. Evidently, the biological assessment of rivers is gradually brought to attention of Chinese river management. Up until the year 2019, no nation-wide assessment program exists, but this is supposed to be changed in the future. The utilized structural metrics for biological quality elements, currently applied mainly for academic purposes, include aquatic macroinvertebrates, fish and benthic algae. For macroinvertebrates the calculation of "family richness", "macroinvertebrate BMWP score" and "EPT family richness" is used. For fish communities the "fish species index", "fish biotic index" and "fish Berger Parker index" are used and for benthic algae the "algal biotic index" and "algal Berger Parker index".

2.1.4.2 Germany

For several years, Germany has been monitoring its running waters according to the demands of the WFD, requiring the assessment of QEs as described below. Hydromorphological and physico-chemical QEs support biological QEs, but cannot replace them. The river status class (Table 2.2) is obtained from a ratio, which represents the relationship between the values of biological parameters observed for a given stream or river and the values of these parameters in the undisturbed reference condition

Table 2.2 General definition of rivers' ecological quality in Germany (adapted from: EC 2000)

Status	Normative definition
High	Physico-chemical/hydromorphological quality elements No, or only very minor, anthropogenic alterations to river type-specific undisturbed reference conditions Biological quality elements Values reflect river type-specific undisturbed reference conditions, and show no, or only very minor, evidence of distortion; river type-specific conditions and communities
Good	Biological quality elements Low levels of distortion resulting from human activity are evident, but deviate only slightly river type-specific undisturbed conditions
Moderate	Biological quality elements Moderate deviations from river type-specific undisturbed reference conditions; moderate signs of distortion resulting from human activity; significantly more disturbed than under good status condition
Poor	Biological quality elements Major deviations from river type-specific undisturbed reference conditions; relevant biological communities deviate substantially from those of river-type specific undisturbed reference conditions
Bad	Biological quality elements Severe alterations from river type-specific undisturbed reference conditions; absence of large proportions of relevant biological communities of river-type specific undisturbed reference conditions

pursuant to the river or stream type. Reference conditions have been defined for 25 stream and river types [51]: four types for the eco-region of the Alps and Alpine foothills, eight types for the Central German Highlands, nine types for the North German Lowlands and four ecoregion-independent types. The worst assessment result of a biological QE determines the overall assessment result (one-out-all-out principle).

Aquatic macroinvertebrates are indicators of local environmental conditions. Since 1991, the method of assessing water quality and the organic impact in running waters based on macroinvertebrates has been regulated by the German standard DIN 38 410 [8]. It is based on one of the traditional saprobic procedures [47]. Samples are taken accordingly to the international standard DIN-EN-ISO 10870 [9]. Calculation, classification and presentation are also standardized. The ecological status can be determined with the assessment system PERLODES [37]. More detailed information for the asessment of running waters using macroinvertebrates as passive indicators is given in [23].

The analysis of macrophytes is a particularly well-suited tool for monitoring long-term trends in trophic conditions [53]. Nationally, the multi-habitat-sampling of macrophytes follows a standardized protocol. Besides coverage, vitality, and sociability of each species, environmental data such as shading, flow velocity, average water depths and substrate type are recorded. In Germany, the assessment of the subcomponent macrophytes is enabled by the assessment system PHYLIB [2]. Alternatively, the NRW-procedure [57] can be used for the assessment of rivers based on macrophytes. Both assessment methods define states of potentially natural vegetation by means of a community approach. The determination of river water quality stems from the occurrence of non-typical vegetation units. PHYLIB is also applied to the subcomponent phytobenthos. Sampling and microscopy follow standardized protocols. During sampling, species coverage and substrate types are recorded. The microscopy protocol requires the determination of taxonomic unit and each unit's frequency of occurrence.

The establishment of a nationwide classification and assessment system of streams using phytoplankton was firstly initiated by the adoption of the WFD. Standardized methods for sampling, handling of samples, and analysis are described in the manual by [40]. It is restricted to medium-sized and major rivers. For the assessment of plankton-dominated streams, the software PhytoFluss is available.

Fish are sensitive to modifications of river morphology and lateral as well as longitudinal river channel connectivity. The survey of fish population characteristics is performed by electrofishing. Six QEs are standardized recorded: species and guilds inventory, species abundance and guilds distribution, age structure, migration (index-based), fish region (index-based), and dominant species (index-based). Those QEs (metrics) are compared to a priori defined river-type reference fish biocenosis which is based on expert knowledge. Depending on the deviation of observed values of QEs from values of the reference, scores are achieved. The average score determines the status classification. The methodology in Germany is based on [15]. The assessment of the subcomponent fish is performed by application of the assessment system fiBS [14].

In contrast to biological elements, hydro-morphological QEs are not primarily decisive for assessing the status of a river, but are determining the distribution as well as habitat heterogeneity, availability and niches for biotic communities. Hydro-morphological QEs of the WFD are morphology, hydrological regime and continuity. The quality element morphology does not only focus on the riverbed (course development, longitudinal profile, bed structure), but also on the banks (cross-section, bank profile) and the floodplain [6]. Rivers are assigned to five structural classes. The allocation to the classes is based on the extent by which a river's morphology deviates from river-type-specific reference condition. A river's hydrology is classified using the parameters connection to groundwater bodies, discharge, and discharge dynamics. On the basis of data and expert knowledge, the intensity of given pressures, such as water abstractions or engineering measures, are related to the potential natural status. The classification into five classes (Table 2.2) is based on the one-out-all-out principle [7]. The procedure is currently undergoing practical trials [20]. A method to classify the continuity of watercourses is being developed by LAWA [20].

According to Annex V of the WFD, five physico-chemical quality elements of running waters need to be monitored: visibility, temperature, oxygen, conductivity, acidification, and nutrient conditions. Water body type-specific background values (very good status) and threshold values (good status/good ecological potential) are listed in Annex 7 to the Ordinance on Surface Waters. Measured values determine the allocation to quality classes. High status is achieved if measured values adhere to river-type-specific background values. If measured values are within a range which guarantees river-type-specific correct functioning, a good status is achieved.

2.2 Methodological Comparison of Fish Community Assessment

As described previously, there are several advantages when biological metrics are complementary considered in surface water health assessment and monitoring. Since fish assemblages are considered to be a highly suitable metric for assessing river health [1, 24, 25], there are ongoing activities in China, predominantly on the academic level, to further develop and adjust fish-based indices for Chinese surface waters. Example applications in China using the "index of biological integrity" (IBI) for river health assessment were reported for the Yangtze River [33], the Liao River [48], and some local small rivers [22, 32]. Structural metrics such as the IBI demand a predetermined sampling strategy and the proper choice of methods to be applied. The choice of efficient sampling techniques is thereby a crucial factor for the success of sampling strategies [5, 41, 45].

The following section of this chapter provides a methodological comparison of fish communities along a 70 km stretch of the Fan River, Lioaning province in China. Specific objectives are the comparison of (i) Active vs. Passive sampling methods, (ii) German vs. Chinese sampling approach and techniques, (iii) Determination of fishing technique effectivity for overall community characterization.

2.2.1 Methodology

2.2.1.1 Choice of Sampling Methods

In contrast to less mobile organisms such as plants and some macroinvertebrates, the assessment of fish communities relies on efficient and non-selective sampling techniques [4, 5]. Since fishing techniques are applied from the early development of human societies on, there is a great variety of different active and passive fishing methods. Today applied standardized methods include electrofishing [27, 41, 49], hook-and-line methods, net-based methods and the use of traps [4, 5, 34, 46, 68]. Catch rates, selectivity patterns and species specificity may differ strongly, however, depending on the sampling method chosen. Thus, an accurate characterization of populations can be difficult [11]. Additionally, the efficiency of the respective fishing method most likely varies with the change of stream size as well as turbidity, current velocity and conductivity [24, 41, 54]. Assessment methods currently applied in

China comprise electrofishing for wadeable [22] or gill-netting for non-wadeable streams [17], as well as casting nets and seining [29]. Standardization of assessment is often not reported or impossible when investigations include catches from local fishermen [29].

2.2.1.2 Area of Experimental Investigation

For the methodological comparison, 6 characteristic sites within the Fan River were chosen together with the Chinese project partners (N 42.10826, E 12.448.517 – N 42.257.50, E 12.360.941). The selection of the sampling area was based on the bilateral workshops held between the German and Chinese partners of the SINOWATER project at the Chair of Aquatic Systems Biology in Freising, Germany in 2016. The Fan River is a tributary to the Liao River, one of the seven major river basins in China. Fan River is a well suited model stream, since within less than 100 km the headwater region and the confluent to the Liao River main stem can be reached. The sampling area covers a gradient from wadeable sections with low-flow conditions at the most upstream site to non-wadeable conditions at the confluent and an elevation gradient from 51 to 370 m above sea level (upper sites dominated by agriculture in the catchment, lower sites with influence of the city of Tieling and water abstraction for irrigation + large reservoir in-between). Site ID and pictures are given in Table 2.3. Sites 3–6 are characterized by a major increased discharge from Zhenziling Reservoir above site 3.

2.2.1.3 Description of Sampling Methods and Design

Herein three active and three passive methods were used for fish community assessment. The survey was conducted in May 2017, at this time of the year typically no extreme flood or drought events occur, which would have hampered sampling activities. Active methods comprised single-pass electrofishing as typically applied in Germany and China as well as seining. As passive methods common minnow traps were used as well as gill-netting and longline-fishing. Specifications of the respective methods are listed in Table 2.4.

The sampling design used in this investigation allowed to simultaneously apply the different fishing methodologies within the same section of the river, including replication of each of the methods and coverage of structural variation. Sampling sections were chosen to represent a wide range of environmental conditions from fast-flowing upstream sections to slow-flowing downstream sections. At each sampling site, a 500 m stretch was selected and divided into three main replicate sections of 150 m each, Within these 150 m sections, there were 30 m assigned for each of the active methods (electrofishing German (EG), electrofishing China (EC) and seining (SN)). Passive fishing methods, longline-fishing (L) and traps (T) were spaced

Table 2.3 Sampling sections for the methodological comparison in Fan River, Liaoning province, China

Sitete 1 - JiaHeChang   Site 2 - BaiQiZhai

Site 3 - XiaoTun   Site 4 - LaoBianTai

Site 5 – FanheBridge (City of Tieling)   Site 6 - HuangHeZi

in between the active fishing sections as well as at the most upstream and most downstream end of the whole sampling site resulting in 5 replicates per method. Gill-netting was only applicable at the most downstream site 6. A graphical representation of the sampling design is given in Fig. 2.1. At the most upstream part a 50 m section for time-based electrofishing was located. For each specimen, the total length (TL) was measured to the nearest 0.1 cm, total weight (WT) was measured with 1 g accuracy on a representative subsample of fish.

Table 2.4 Specification of applied active and passive fishing methods

Method type	Specification	
Active		
Electrofishing German (EG)	Graßl ELT62, 3 kW (300 V) engine-powered backpack-electrofisher connected to 1 handheld anode with a 30cm dipnet and a passive braided wire copper cathode. Fish were collected by an assistant person with a dipnet and transferred to a bucket.	
Electrofishing Chinese (EC)	Two copper electrodes on telescope handles, one equipped with a 20 cm dipnet, powered by a 24V battery. Fish were collected by the operator and collected by a second person carrying a bucket.	
Electrofishing Chinese - time based (EC30)	Same equipment as EC. Sampling was conducted by fishing a zigzag route across the sampling section for 30 minutes.	
Seining	15 m net, mesh-size 10 mm, lead-bottomline. Spreading of the net in 90° angle to the river bank towards the middle of the river, then pushing the net back in a half-cirlce towards the bank	
Passive		
Trapping (T)	Minnow trap, diameter 30cm, length 60 cm. Commercial cat-food used as bait. Placed in deeper sections with the whole trap being submerged.	
Longline (L)	5 m fishing line -3 types barbed hooks (#1, #6, #10) replicated 5 times. Each hook size was equipped with different bait: fish filet, mussel meat or worms, and maggot.	
Gillnet	20 m long, 1 m wide, 2 cm mesh. Brought into a stagnant deep area by boat.	

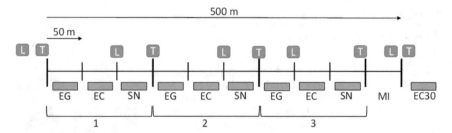

Fig. 2.1 Schematic representation of the sampling design applied. EG: Electrofishing German, EC: Electrofishing China, EC30: 30 min Electrofishing china, SN: Seining, L: Longline fishing, T: Trapping, 1–3: replicate sections, MI: additional zone for macroinvertebrate sampling

2.2.1.4 Calculated Metrics

For comparison of the method-dependent catch, the following metrics are reported: Species abundance, Species richness (d), Shannon diversity index (H'), Eveness (J') Condition indices for each species (BMI), species biomass. To express the effectiveness of each method, the Catch per Unit Effort (CPUE) was calculated. For comparability the CPUE per 10 min is presented. Richness was calculated by counting all the taxa for each site, specific calculations are given below.

$$H = -\sum_{i=1}^{S} P_i \log_2 Pi \tag{2.1}$$

where S is the total number of taxa, Pi is the relative abundance of species i.

$$BMI = \frac{(W_T \times 100)}{L_T^3} \tag{2.2}$$

WT: Total weight (nearest g); LT: Total length (nearest 0.1 cm).

$$CPUE = \frac{N_i}{t_i} \tag{2.3}$$

Ni: Number of individuals per replicate section i; ti: time of fishing at replicate section i

$$d = \frac{S-1}{\log_e N} \tag{2.4}$$

$$J' = \frac{H'}{\log_e S} \tag{2.5}$$

$$1 - \lambda' = 1 - \frac{\{\Sigma i \, Ni(Ni-1)\}}{\{N(N-1)\}} \tag{2.6}$$

S: Total number of taxa; N: Total number of individuals.

2.2.2 Results

2.2.2.1 Fish Community

A total of 33 fish species from 10 different families comprising 7799 specimen were collected by all methods (and sites) combined. Most of the sampling sites within the Fan River system showed to be dominated by Cyprinidae, comprising up to 85% of all fish specimen caught, followed by Nemacheilidae, Cobitidae and Gobiidae. The fish community was generally characterized by generalists like *Rhynchocypris lagowskii*, *Barbatula nuda* and *Zacco platypus*. These first two species represented 21.9% and 23.7% of the total fish biomass, respectively. They were found at every site except the most downstream site 6, HuangHeZi, where the fish assemblage was dominated by *Carassius auratus* and *Abbottina rivularis*, most likely due to the vicinity of the Liao River main stem. In general, no apex predators or fish larger 16 cm (total length) were caught by the methods applied or are simply missing in Fan River. A full list of the Fan River fish community is presented in Table 2.5. *Cobitis granoei*, *Pseudorasbora parva* and *Zacco platypus* were detected at all sites. Some species such as *Aphyocypris chinensis* (site 2), *Chanodichthys erythropterus* (site 6), *Hypomesus olidus* (site 6), *Opsariichthys bidens* (site 2), *Romanogobio tenuicorpus* (site 6) were only detected at single sites. In site 6 three of the species exclusively occurred, indicating a strong influence of the Liao River on the fish community.

For 9 species less than 10 individuals were found (*Chanodichthys erythropterus*, *Gymnogobius urotaenia*, *Tachysurus fulvidraco*, *Hypomesus olidus*, *Pungitius sinensis*, *Misgurnus mohoity*, *Silurus asotus*, *Oryzias sinensis*, *Opsariichthys bidens*).

In terms of metrics the German electrofishing method resulted in the highest value for α-diversity with H' = 2.35 in contrast to seining (1.65), Table 2.6. The catch from trapping resulted in the highest values for species richness with d = 3.936.

2.2.2.2 Method Efficiency

In comparison of overall method efficiency, none of the methods turned out to be suitable to cover the full species range detected at each site compared to all methods combined. The electrofishing methods, however, detected always at least >50% of the species, while seining detected less than 40%. The results using the German electrofishing method (EG) had the greatest overall coverage in terms of number of taxa and biomass, and was suited to detect 56–100% of the species present at the specific sites. This result is further corroborated by the comparison of the catch per unit effort (CPUE). The electrofishing methods in general was the most effective method to characterize fish assemblages. The calculated CPUE generally revealed

Table 2.5 List of fish species in the Fan River system. Total number of individuals (N), percentage of total catch (Proportion) and biomass is given

Species name	Sum N	Proportion [%]	Biomass [kg]
Rhynchocypris lagowskii	2484	31.8	4.56
Barbatula nuda	1683	21.6	3.00
Abbottina rivularis	481	6.17	1.49
Carassius auratus	377	4.8	4.94
Cobitis granoei	374	4.8	0.39
Pseudorasbora parva	340	4.4	0.66
Rhinogobius giurinus	337	4.3	0.52
Misgurnus anguillicaudatus	298	3.8	0.58
Rhinogobius brunneus	218	2.8	0.14
Zacco platypus	195	2.5	0.52
Rhinogobius cliffordpopei	194	2.5	0.22
Rhodeus lighti	187	2.4	0.19
Microphysogobio chinssuensis	164	2.1	0.27
Gobio cynocephalus	107	1.4	0.31
Acanthorhodeus chankaensis	91	1.2	0.22
Micropercops swinhonis	60	0.8	0.10
Paramisgurnus dabryanus	55	0.7	0.15
Aphyocypris chinensis	29	0.4	0.05
Romanogobio tenuicorpus	23	0.3	0.03
Rhodeus lighti	21	0.3	0.02
Hemiculter leucisculus	20	0.3	0.01
Lefua costata	17	0.2	0.02
Oryzias sinensis	9	0.1	< 0.01
Opsariichthys bidens	9	0.1	0.44
Silurus asotus	8	0.1	0.31

(continued)

Table 2.5 (continued)

Species name	Sum N	Proportion [%]	Biomass [kg]
Misgurnus mohoity	5	<0.1	0.03
Pungitius sinensis	4	<0.1	<0.01
Hypomesus olidus	3	<0.1	<0.01
Gymnogobius urotaenia	2	<0.1	<0.01
Tachysurus fulvidraco	2	<0.1	0.24
Chanodichthys erythropterus	1	<0.1	0.05

Table 2.6 Diversity indices of fish community in Fan River based on the result of the different sampling methods

Method	Total species S	Total individuals N	Species richness D	Eveness J'	Shannon diversity H'	Simpson diversity 1-Lambda
EC	24	1954	3.03	0.67	2.15	0.81
EC 30	27	1714	3.49	0.52	1.74	0.64
EG	28	3570	3.30	0.70	2.35	0.84
SN	11	125	2.07	0.69	1.65	0.69
T	24	345	3.93	0.74	2.37	0.83
L	0	0	-	-	-	-

significant differences between all methods (KW, $H(3) = 33.7$, $p < 0.001$)[1]. Within the active methods, the seining resulted in a significant lower CPUE compared to electrofishing methods ($p < 0.05$) and the passive method trapping showed significant lower CPUE compared to each other method ($p < 0.001$). Seining indicated to be the least efficient method of the active methodologies, however still yielded a higher CPUE compared to passive methods. Longline fishing never yielded any catch, regardless the site in which this method was applied. The method efficiency comparison is presented in Table 2.7.

Gill-netting was not applicable in the upstream parts of the river with higher flow-velocities and shallow water depths. In the most downstream, deep and low flowing reaches, results from the gill-net differed in terms of species composition from all the other methods applied, presumably caused by being more selective towards pelagic species such as *Carassius auratus* or *Hypomesus olidus* which were hardly caught with the active fishing methods applied from the river bank as used in this comparison.

---

[1] Statistical comparisons to test for differences between groups were performed by using the Kruskal–Wallis rank-sum test, with Mann–Whitney U pairwise comparisons (Bonferroni corrected). Significance was accepted at $p < 0.05$.

Table 2.7 Method efficiency comparison. Total number of taxa detected for each method and site. Values in brackets represent percentage of taxa detected in relation to total number of taxa by all methods combined

Site	N species	EC	EC_{30}	EG	G	L	SN	T
1	9	5 (56%)	7 (78%)	5 (56%)	N/A	0	2 (22%)	2 (22%)
2	19	15 (79%)	16 (84%)	16 (84%)	N/A	0	5 (26%)	10 (53%)
3	15	13 (87%)	11 (73%)	15 (100%)	N/A	0	5 (33%)	11 (73%)
4	16	11 (69%)	11 (69%)	13 (81%)	N/A	0	0 (0%)	10 (63%)
5	19	10 (53%)	12 (63%)	17 (89%)	N/A	0	7 (37%)	9 (47%)
6	21	9 (42%)	12 (57%)	15 (71%)	11 (52%)	0	5 (24%)	6 (29%)

Some of the species that were only detected by method EG comprise bottom-dwelling species or "rare" species with overall specimen <10 at all sites combined.

2.3 Discussion

The results of this study highlight potential benefits of integrating standardized biological monitoring tools of fish community assessments that can be combined with existing Chinese standards for physico-chemical metrics used to rank surface water bodies into water quality grades I to V+. The grading system stems from the perspective of water usage purposes, such as suitability for drinking or irrigation purposes. This is in contrast to the European system that primarily focusses on the alterations of the ecological status of its aquatic systems including the multiple impacts on them. With respect to the recent rapid economic development, impacts on aquatic ecosystems in China have heavily increased. In the year 2014 overall water quality did not meet the criteria for the general public and fishery use (at least Grade III) from 28.8% of the monitoring sections in major rivers [66]. In 2017, the physico-chemical water quality of rivers showed for 23.8% of the monitoring sections a quality grade IV and V, while 8.3% failed to meet the grade V standard [38]. Investigations by Chinese researchers have previously identified physical habitat alterations and chemical pollutants as key factors influencing fish communities in the Liao River Basin [17, 65]. They also reported that, compared with the historical data, almost half of the fish species have disappeared and some local species were replaced by invasive ones. To meet the goal to restore biological integrity in the Liao River system and other watersheds in China, it becomes obvious that monitoring tools should consider biological quality elements such as fish. Since biomonitoring methods are already regionally tested, it can be anticipated that China's national assessment approach changes in the near future and suitable protocols will be developed soon. European monitoring programs (including Germany) comprise various types of quality elements. Biological water QEs are thereby primarily decisive for assessing a river's actual status. Physico-chemical and hydromorphological QEs support the biological QEs. Hereby, a large spectrum of the complexity of different biological and ecological characteristics as well as the large variety of pressures and impacts can be covered. One strength of this approach is the general comparability over space and time, which allows to detect deviations between the observed conditions and the respective reference conditions. Specific metrics, as applied in the European WFD cannot be used for a direct comparison here. For Chinese Rivers there is so far limited information on guilds distribution, migration (index-based), fish region (index-based), and dominant species (index-based) as used in Europe.

For a representative assessment not only the choice of methods is crucial, but also the consideration of site-specific characteristics [4, 5, 41]. Depending on present fish communities either passive or active bioindication methods [45] might be suitable, ideally comprising a combination of different methods. In the method comparison presented here, the differences between the applied sampling strategies could exemplarily be shown. However, no method was suitable to detect all the species present at a given site. Only a combination of active and passive sampling techniques provides a best-case coverage. Electrofishing-based methods showed to be most effective, regarding range, efficiency and the potential to detect rare species. In this direct comparison the German electrofishing method revealed to be more suitable in comparison to the other applied methods in terms of efficiency and CPUE. In general it can be stated that active methods such as electrofishing should preferentially be used for fish community assessment in running waters [10]. There were minor differences between the Chinese and the German electrofishing, however, most likely this is rather a function of the established voltage by the different power sources used. Ideally the power of the electrofishing gear is adequately chosen, based on the river-specific size, discharge, depth and conductivity [41]. This approach is already applied in European fish stock assessments like the European Water Framework directive monitoring programs [10]. For headwater areas and small wadeable rivers, the Chinese electrofishing gear proved to be sufficient to characterize the present fish community. However, when species richness (or age structure) is the target of the fish assessment, this method increases the risk of missing species (or size classes). Thereby the time-based approach EC30 provided a broader spectrum compared to the single-pass fishing approach. In contrast, the estimation of the ecological status does not require the determination of absolute fish densities, but relative species composition and abundance [58]. During electric fish sampling, species composition and relative abundance are likely to be biased because of differences in the catchability of different species and fish sizes [3, 11, 56], which demands sufficient efficiency of the methods. Based on the difficulty of detecting rare species, random sampling should be conducted in conjunction with targeted sampling of rare species or passive methods such as trapping (see also [55]). Methods such as the Longline fishing showed to be not suitable for fish community characterization in this study.

No historic data is available for the Fan River, however, the increasing anthropogenic pressure and resulting aquatic species decline is indicated by many experts. This highlights the overall value of biological monitoring for aquatic health status assessment. More detailed information on biological integrity is urgently needed, especially for monitoring of the future status of these river systems.

References

1. Angermeier PL, Karr JR (1986) Applying an index of biotic integrity based on stream-fish communities: considerations in sampling and interpretation. N Am J Fish Manag 6:418–429. https://doi.org/10.1577/1548-8659(1986)6%3c418:AAIOBI%3e2.0.CO;2
2. Bayerisches Landesamt für Umwelt, LfU (2005) Bewertungsverfahren Makrophyten and Phytobenthos Fließgewässer- und Seen-Bewertung in Deutschland nach EG-WRRL, Informationsheft 1/05. LfU München, Germany
3. Bohlin T, Hamrin S, Heggberget TG, Rasmussen G, Saltveit SJ (1989) Electrofishing–theory and practice with special emphasis on salmonids. Hydrobiologia 173:9–43. https://doi.org/10.1007/BF00008596
4. Bonar SA, Contreras-Balderas S, Iles AC (2009) Introduction to standardized sampling. In: Bonar SA, Hubert WA, Willis DW (eds) Standard methods for sampling North American freshwater fishes. American Fisheries Society, Bethesda, pp 1–12
5. Brandner J, Pander J, Mueller M, Cerwenka A, Geist J (2013) Effects of sampling techniques on population assessment of invasive round goby. J Fish Biol 82:2063–2079. https://doi.org/10.1111/jfb.12137
6. Bund/Länder-Arbeitsgemeinschaft Wasser, LAWA (2011). Überarbeitung der Verfahrensbeschreibung der Gewässerstrukturkartierung in der Bundesrepublik Deutschland. Verfahren für kleine bis mittelgroße Fließgewässer. Berlin, Germany: Kulturbuchverlag GmbH
7. Bund/Länder-Arbeitsgemeinschaft Wasser, LAWA (2014). Klassifizierung des Wasserhaushalts von Einzugsgebieten und Wasserkörpern – Verfahrensempfehlung. a) Handlungsanleitung. https://www.gewaesser-bewertung.de/files/lawa_wh_verfahrensempfehlung.pdf
8. DIN EN ISO 38410-1. German standard methods for the examination of water, waste water and sludge - Biological-ecological analysis of water (group M) - Part 1: Determination of the saprobic index in running waters (M 1)
9. DIN EN ISO 10870. Water quality - Guidelines for the selection of sampling methods and devices for benthic macroinvertebrates in fresh waters (ISO 10870:2012); German version EN ISO 10870:2012
10. CEN EN 14011 8 (2003) Water quality–sampling of fish with electricity, 18 pp. European Committee for Standardization, Brussels
11. Daulwater DC, Fisher WL (2007) Electrofishing capture probability of smallmouth bass in streams. N Am J Fish Manag 27:162–171. https://doi.org/10.1577/M06-008.1
12. DIN EN ISO 10870 Water quality - Guidelines for the selection of sampling methods and devices for benthic macroinvertebrates in fresh waters (ISO 10870:2012); German version EN ISO 10870:2012
13. Dudgeon D et al (2006) Freshwater biodiversity: importance, threats, status and conservation challenges. Biol Rev 81(2):163–182. https://doi.org/10.1017/S1464793105006950
14. Dußling U (2009) Handbuch zu fiBS–Schriftenreihe des Verbandes Deutscher Fischereiverwaltungsbeamter und Fischereiwissenschaftler e.V., Heft 15. www.gewaesser-bewertung.de/files/fibs-handbuch_2009.pdf
15. Dußling U, Berg R, Klinger H, Wolter C (2004) Assessing the ecological status of river systems using fish assemblages. Handbuch Angewandte Limnologie 20. Erg.Lfg. 12/04, pp 1–84
16. European Parliament and Council of the European Union (2000) Directive 2000/60/EC of the European Parliament and the Council of 23 October 2000 establishing a framework for Community action in the field of water policy. https://eur-lex.europa.eu/legal-content/EN/TXT/?uri=CELEX:32000L0060

17. Gao X, Zhang Y, Ding S, Zhao R, Meng W (2015) Response of fish communities to environmental changes in an agriculturally dominated watershed (Liao River Basin) in northeastern China. Ecol Eng 76:130–141. http://dx.doi.org/10.1016/j.ecoleng.2014.04.019
18. GB 5749-2006 Standards for Drinking Water Quality (2006) National Standard of the People's Republic of China. Ministry of Health
19. Geist J (2015) Seven steps towards improving freshwater conservation. Aquat Conserv 25:447–453. https://doi.org/10.1002/aqc.2576
20. German Environment Agency, UBA (2017) Waters in Germany: status and assessment. Dessau-Roßlau, Germany: UBA. https://www.umweltbundesamt.de/publikationen/waters-in-germany
21. Jähnig SC, Qinghua C (2010) River water quality assessment in selected Yangtze tributaries: background and method development. J Earth Sci 21(6):876–881. https://doi.org/10.1007/s12583-010-0140-y
22. Jia YT, Sui XY, Chen YF (2013) Development of a fish-based index of biotic integrity for wadeable streams in Southern China. Environ Manag 52:995–1008. https://doi.org/10.1007/s00267-013-0129-2
23. Johnson RK, Wiederholm T, Rosenberg DM (1993) Freshwater biomonitoring using individual organisms, populations, and species assemblages of benthic macroinvertebrates. In: Rosenberg DM, Resh VH (eds) Freshwater biomonitoring and benthic macroinvertebrates, pp 40–158. Chapman & Hall, New York
24. Karr JR (1981) Assessment of biotic integrity using fish communities. Fisheries 6:21–27
25. Karr JR, Chu EW (1999) Restoring life in running waters: better biological monitoring. Island Press, Washington
26. Kummu M, De Moel H, Porkka M, Siebert S, Varis O, Ward PJ (2002) Lost food, wasted resources: global food supply chain losses and their impacts on freshwater, cropland, and fertiliser use. Sci Total Environ 438:477–489. https://doi.org/10.1016/j.scitotenv.2012.08.092
27. Lapointe NWR, Corkum LD, Mandrak NE (2006) Point sampling by boat electrofishing: a test of the effort required to assess fish communities. N Am J Fish Manag 26:793–799. https://doi.org/10.1577/M06-007.1
28. Li F, Cai Q, Ye L (2010) Developing a benthic index of biological integrity and some relationships to environmental factors in subtropical Xiangxi river. China. Int Rev Hydrobiol 95(2):171–189. https://doi.org/10.1002/iroh.200911212
29. Li T, Huang X, Jiang X, Wang X (2018) Assessment of ecosystem health of the Yellow River with fish index of biotic integrity. Hydrobiologia 814:31–43. https://doi.org/10.1007/s10750-015-2541-5
30. Li G-Y (2004) Keeping healthy life of the yellow river - an ultimate aim of taming the Yellow river. Yellow River 26:1–3
31. Liess M, Von der Ohe PC (2005) Analyzing effects of pesticides on invertebrate communities in streams. Environ Toxicol Chem 24:954–965. https://doi.org/10.1897/03-652.1
32. Liu K, Zhou W, Li FL, Lan JH (2010) A fish-based biotic integrity index selection for rivers in Hechi prefecture, Guangxi and their environmental quality assessment. Dongwuxue Yanjiu 31(5):531–538. https://doi.org/10.3724/SP.J.1141.2010.05531
33. Liu MD, Chen DQ, Duan XB, Wang K, Liu SP (2009) Assessment of ecosystem health of Upper and Middle Yangtze river using fish-index of biotic integrity. J Yangtze River Sci Res Inst 27(2):1–6
34. Lynch PL, Mensinger AF (2011). Seasonal abundance and movement of the invasive round goby (Neogobius melanostomus) on rocky substrate in the Duluth–Superior Harbor of Lake Superior. Ecol Freshw Fish 21:64–74. https://doi.org/10.1111/j.1600-0633.2011.00524.x
35. Maloney EM (2019) How do we take the pulse of an aquatic ecosystem? Current and historical approaches to measuring ecosystem integrity. Environ Toxicol Chem 38(2):289–301. https://doi.org/10.1002/etc.4308
36. McRae L, Deinet S, Freeman R (2017) The diversity-weighted living planet index: controlling for taxonomic bias in a global biodiversity indicator. PLoS ONE 12:1–21. https://doi.org/10.1371/journal.pone.0169156

37. Meier C, Haase P, Rolauffs P, Schindehütte K, Schöll F, Sundermann A, Hering D (2006) Methodisches Handbuch Fließgewässerbewertung-Handbuch zur Untersuchung und Bewertung von Fließgewässern auf der Basis des Makrozoobenthos vor dem Hintergrund der EG-Wasserrahmenrichtlinie. Methodical stream assessment manual on the basis of the macrozoobenthos and with regards to the European water frame directive

38. Ministry of Ecology and Environment of the People's Republic of China, MEP (2018) 2017 report on the state of the ecology and environment in China. http://english.mee.gov.cn/Resour ces/Reports/soe/

39. Ministry of Environmental Protection of the People's Republic of China, MEP (2002) Environmental quality standards for surface water (GB3838–2002). https://www.chinesestandard. net/PDF/English.aspx/GB3838-2002

40. Mischke U, Behrendt H (2007) Handbuch zum Bewertungsverfahren von Fließgewässern mittels Phytoplankton zur Umsetzung der EU-WRRL in Deutschland. Schweizerbart, Stuttgart

41. Mueller M, Pander J, Knott J, Geist J (2017) Comparison of nine different methods to assess fish communities in lentic flood-plain habitats. J Fish Biol 91:144–174. https://doi.org/10.1111/ jfb.13333

42. National People's Congress, NPC (2016). The 13th Five Five-Year-Plan for the Economic and Social Development of the People's Republic of China 2016–2020. Beijing, China: Central Compilation & Translation Press

43. Palmer MA, Febria CM (2012) The heartbeat of ecosystems. Science 336:1393–1394. https:// doi.org/10.1126/science.1223250

44. Pan B et al (2013) An exploratory analysis of benthic macroinvertebrates as indicators of the ecological status of the Upper Yellow and Yangtze Rivers. J Geogr Sci 23(5):871–882. https:// doi.org/10.1007/s11442-013-1050-6

45. Pander J, Geist J (2013) Ecological indicators for stream restoration success. Ecol Indic 30:106– 118. https://doi.org/10.1016/j.ecolind.2013.01.039

46. Pander J, Mueller M, Knott J, Geist J (2018) Catch-related fish injury and catch efficiency of stow-net-based fish recovery installations for fish-monitoring at hydropower plants. Fish Manag Ecol 25:31–43. https://doi.org/10.1111/fme.12263

47. Pantle R, Buck H (1955) Die Biologische Überwachung der Gewässer und die Darstellung der Ergebnisse. Besondere Mitteilungen zum Gewässerkundlichen Jahrbuch 12:159–162

48. Pei XJ, Niu CJ, Gao X, Xu C (2010) The ecological health assessment of Liao River Basin, China, based on biotic integrity index of fish. Shengtai Xuebao/Acta Ecologica Sinica 30(21):5736–5746

49. Persat H, Copp GH (1990) Electric fishing and point abundance sampling for the ichthyology of large rivers. In: Cowx IG (ed) Developments in Electric Fishing. Kluwer, Amsterdam, pp 197–209

50. Pignata C et al (2013) Application of European Biomonitoring techniques in China: are they a useful tool? Ecol Indic 29:489–500. https://doi.org/10.1016/j.ecolind.2013.01.024

51. Pottgiesser T, Sommerhäuser M (2008) Beschreibung und Bewertung der deutschen Fließgewässertypen-Steckbriefe und Anhang. https://gewaesser-bewertung.de/files/steckb riefe_anhang_april2008.pdf

52. Rapport DJ, Costanza R, McMichael AJ (1998) Assessing ecosystem health. Trends Ecol Evol 13:397–402. https://doi.org/10.1016/S0169-5347(98)01449-9

53. Schneider S, Melzer A (2003) The trophic index of macrophytes (TIM): a new tool for indicating the trophic state of running waters. Int Rev Hydrobiol 88(1):49–67

54. Simon TP, Sanders RE (1999) Applying an index of biotic integrity based on great river fish communities: considerations in sampling and interpretation. In: Simon TP (ed) Assessment approaches for estimating biological integrity using fish assemblages. Florida, Boca Raton, pp 475–505

55. Smith KL, Jones ML (2005) Watershed-level sampling effort requirements for determining riverine species composition. Can J Fish Aquati Sci 62:1580–1588. https://doi.org/10.1139/ f05-098

56. Speas DW, Walters CJ, Ward DL, Rogers RS (2004) Effects of intraspecific density and environmental variables on electrofishing catchability of brown and rainbow trout in the Colorado River. N Am J Fish Manag 24:586–596. https://doi.org/10.1577/M02-193.1
57. Van de Weyer K (2015) NRW-Verfahren zur Bewertung von Fließgewässern mit Makrophyten. Fortschreibung und Metrifizierung, LANUV-Arbeitsblatt 30. LANUV, Recklinghausen, Germany
58. Vehanen T, Sutela T, Jounela P, Huusko A, Mäki-Petäys A (2013) Assessing electric fishing sampling effort to estimate stream fish assemblage attributes. Fish Manag Ecol 20:10–20. https://doi.org/10.1111/j.1365-2400.2012.00859.x
59. Vinebrooke RD, Cottingham KL, Norberg J, Scheffer M, Dodson SI, Maberly SC, Sommer U (2004) Impacts of multiple stressors on biodiversity and ecosystem functioning: the role of species co-tolerance. Oikos 104:451–457. https://doi.org/10.1111/j.0030-1299.2004.13255.x
60. Vitousek PM, Mooney HA, Lubchenco J, Melillo JM (1997) Human domination of earth's ecosystems. Science 277:494–499. https://doi.org/10.1126/science.277.5325.494
61. Vörösmarty CJ et al (2010) Global threats to human water security and river biodiversity. Nature 467:555–561. https://doi.org/10.1038/nature09440
62. Wang L, Gong Z, Zhang J, Li Y (2007) Comprehensive index system for evaluation of river ecosystem health. China Water Wastewater 22(23):97–100
63. Wang Q, Yuan X, Liu H, Xu PX, Wang Z, Zhang Y (2014) Stream Habitat assessment of Dong River, China, using River Habitat survey method. Sheng Tai Xue Bao/Acta Ecologica Sinica 34(6):1548–1558. https://doi.org/10.5846/stxb201210201458
64. Wang Y et al (2014) A national pilot scheme for monitoring of ecological integrity of surface waters in China. Environ Dev 10(1):104–107. https://doi.org/10.1016/j.envdev.2014.02.003
65. Xie YH (2007) Freshwater fishes in Northeast Region of China. Liaoning Science and Technology Press, Shengyang, p 529
66. Xing Y, Zhang C, Fan E, Zhao Y (2015) Freshwater fishes of China: species richness, endemism, threatened species and conservation. Divers Distrib 22(3):358–370. https://doi.org/10.1111/ddi.12399
67. Yang L-F, Li Y-W, Qi D-G et al. (1992) The assessment of aquatic insect community structure and biological water quality in Jiuhua River. Acta Ecol Sinica 12(1):8–15
68. Young JAM, Marentette JR, Gross C, McDonald JI, Verma A, Marsh-Rollo SE, Macdonald PDM, Earn DJD, Balshine S (2010) Demography and substrate affinity of the round goby (Neogobius melanostomus) in Hamilton Harbour. J Great Lakes Res 36:115–122. https://doi.org/10.1016/j.jglr.2009.11.001
69. Yue Z-M (2005) Maintenance of the Pearl River health life, building a Green Pearl River, and Strive to build a Harmonious society. Pearl River 2:1–5. https://doi.org/10.3969/j.issn.1001-9235.2005.03.001
70. Zhao Y, Yang Z (2005) Preliminary study on assessment of urban river ecosystem health. Adv Water Sci 16(3):349–355

Chapter 3
Dianchi Shallow Lake Management

Florian Rankenhohn, Tido Strauß, and Paul Wermter

3.1 Introduction

3.1.1 Conceptual Thoughts Connecting Water Management and Shallow Lake Management

The SINOWATER-Project funded by Federal Ministry of Education and Research of Germany (BMBF) examined and discussed the idea of an Integrated Water Resources Management on all levels of water management. As shown in Fig. 3.1 this covers a range from the overall management principles of Good Water Governance (Grambow [7], Disse and Chen [3]) over conceptual levels to technical solutions like stormwater management (Wang et al. [35]). This chapter is dedicated to the conceptual level of modern water management as shown in the following figure. Measurement planning on this conceptual level benefits from strong sustainable management principles (Grambow [7]). Within the Kunming part of the SINOWATER-Project, we also committed a major part of our work towards technical solutions, side by side with our colleagues from Kunming authorities. Included in the book you will find a thorough

F. Rankenhohn · P. Wermter
Research Institute for Water and Waste Management at RWTH Aachen (FiW), Kackertstr. 15-17, 52056 Aachen, Germany

F. Rankenhohn (✉)
Department of Civil Engineering, University of Kaiserslautern, Paul-Ehrlich Str. 14, 67663 Kaiserslautern, Germany
e-mail: florian.rankenhohn@bauing.uni-kl.de

T. Strauß
Research Institute for Ecosystem Analysis and Assessment (Gaiac), Kackertstr. 10, 52072 Aachen, Germany

P. Wermter
Ministry of Climate Protection, Environment, Energy and Mobility of Rhineland-Palatinate, Kaiser-Friedrich-Str. 1, 55116 Mainz, Germany

© The Author(s) 2022
M. Dohmann et al. (eds.), *Chinese Water Systems*, Terrestrial Environmental Sciences,
https://doi.org/10.1007/978-3-030-80234-9_3

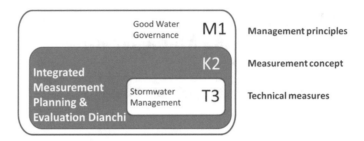

Fig. 3.1 Levels of water management implemented as sub-projects M1, K2 and T3 in BMBF-funded SINOWATER-project

description of measures of storm water management planned and implemented in the tributaries of the Caohai and its urban river basin (Wang et al. [35]).

In this chapter, we would like to introduce our readers to the outcomes of our studies along the northern part of Dianchi the Caohai at Kunming, the capital of Yunnan Province. Dianchi is the third biggest Chinese shallow lake. This lake is under constant pressure of the growth of the city.

3.1.2 Two Studies for One Goal

The following chapters describe some of the SINOWATER-Project contributions to the shallow lake management strategies of the Kunming Dianchi Management Authorities. These recommendations are based on two different studies. The first study served to collect data and to broaden the database by taking and analysing additional water and sediment samples. The second study comprised the shallow lake modelling and discussed some of the six above mentioned management scenarios on a scientific basis. The analysis of the studies are following the overall idea of the combined approach of EU-Water Framework Directive. It is the combination of basic technical emission control and strict enforcement of environmental quality standards (EQS). In consequence, the EQS enforcement is a quality management step aiming at checking if technical emission control is sufficient to sustain the targeted water quality. If necessary because basic technical control measures are insufficient, the evaluation of the EQS-Check will lead to an extension of quality control measures. This water management approach has been enforced in the EU with the introduction of EU-Water Framework Directive in the year 2000.

In recent years, numbers of Kunming Academy of Environmental Sciences and other Dianchi Management Authorities show a decrease in lake pollution due to the strong commitment of environmental protection. We wish our partners and colleagues strong perseverance and the best support and full confidence from the political leadership for a permanent journey together.

Worldwide a lot of shallow lakes suffer from bad water quality caused by eutrophication (Küppers et al. [21]; Qinghui et al. [27]; Diovisalvi et al. [6]; Lei et al. [19]; Hargeby et al. [10]; Vadeboncoeur et al. [33]). In China, bad water quality and regular algae blooms in shallow lakes such as Taihu, Chaohu and Dianchi are a huge environmental problem, caused by high nutrient intakes due to growing population, industrial growth and high fertilizer inputs from agricultural sides (Nixdorf & Zhou [25]; Zhang et al. [50]; Huang et al. [13]; Cao et al. [2]; Li et al. [46]; Qin et al. [26]; Zhu et al. [52]). All three lakes are major national water resources of China. They are serving not only as habitat for flora and fauna but also perform tasks in areas such as drinking water supply, local recreation and tourism. For shallow lakes once the low carrying capacity against pollution is exceeded the negative effects such as eutrophication with extreme algae bloom, fish mortality, water toxicity and others can cause severe effects for the regional water management and put monetary pressure on regional or even provincial economy.

Lake Dianchi in the Chinese province Yunnan is a shallow lake suffering from algae blooms for years due to high nutrient intakes (Zhang et al. [51]; Wu et al. [40]; Wang et al. [38]; Jin [16]). Its capital Kunming, which is next to the lake, has grown from 3.7 million inhabitants in 1978 to 6.8 million in 2017 (Kunming Statistics Bureau [18]). Meanwhile, the bad water quality causes a negative feedback on the development of the city of Kunming (Yang [45]). In recent years, local authorities built several wastewater treatment plants and conducted projects to improve the water quality of Lake Dianchi. Nevertheless, there are still algae outbreaks, especially in the summer monsoon season.

The most polluted part of is Lake Dianchi its northern part, Caohai (Zhan et al. [48]). Separated from the southern part Waihai, it receives huge amounts of wastewater from the city of Kunming. In the past, Lake Caohai was scarcely sampled, ignoring the fact that the Caohai-Waihai Lock separates Lake Caohai and Lake Waihai since 1996 (Yan et al. [43]). For that reason, Lake Caohai had to be investigated as a water system of its own, which we did in this study that was part of the subproject Lake Management in the SINOWATER project.

In this study the aim was to sample Lake Caohai over the summer monsoon period in a holistic way like proposed from (Yua et al. [47]). Thus, we took water samples from all seven tributaries and from the Lake itself at three dates and analysed them on nutrients providing algae growth. As it is known that nutrient release from sediments can have a huge impact on the water quality (Moore et al. [23]), we took also sediment samples from Lake Caohai. With these, we performed nutrient release tests. Some of the generated results were the basis for a numerical simulation with the stoichiometric lake model StoLaM (see Chap. 3.3).

3.2 Study 1: Water and Sediment Monitoring Data Acquisition

3.2.1 Research Area

Sampling was carried out on the northern part of the Dianchi. This section, called Caohai or Inner Lake, is located directly at the city of Kunming. It is 8.15 km^2 in size, which corresponds to 2.7% of the total area of the Dianchi and accounts for about 1% of the total volume (Jin et al. [16]). I was separated from the larger Waihai in 1996 (He et al. [12]).

In recent years, algae blooms have occurred regularly in Caohai, with cyanobacteria playing a particularly important role (Han et al. [9]; Yua et al. [47]). The Caohai receives 45% of the total amount of wastewater flowing into the Dianchi and shows significantly increased nitrogen and phosphorus concentrations (Wang et al. [39]; Zhan et al. 2016). The Caohai is divided into two larger parts by the Dongfengba Dam. This division was carried out in order to retain the heavily polluted water from the Wangjiadui, Xinyunliang and Laoyunliang rivers in the western part and to purify it mechanically by sedimentation. It is fed into the artificial lake outlet, the western tunnel, thus separating the eastern part from this source of pollution (Yan et al. [43]). The rivers Wulong, Daguan, Xiba and Chuanfang flow into the eastern part. Furthermore, in the southern part of the Caohai there is another separated area that is used for research on aquatic plants. The western tunnel forms the Caohai's outflow and diverts the water at Anning into the Tanglang River, which belongs to the Yangtze River catchment area (Jin et al. [16]).

The Wangjiadui is an outflow from an electric power station, which is no longer fed and has only a low water flow due to rainwater. It is part of several smaller inflows at the West Bank of the Caohai (He et al. [12]).

The Xinyunliang River has its source in the mountains north of Kunming City and flows for about 21 km, first through the more rural and then through the more densely populated areas of Kunming City. It is fed by 15 smaller rivers (Nie et al. [24]). He et al. [12] reported that around 40% of the discharge are fed by WWTP. The Xinyunliang River is described as the most polluted tributary to the Dianchi and its main source of pollution (Wang et al. [36]; Huang et al. [15]). The Xinyunliang contains 41.28% of the industrial wastewater of the entire Dianchi catchment area (Nie et al. [24]). Since the construction of a wastewater treatment plant along the river and individual projects to improve the water quality of the river, the river's fauna has recovered considerably (Huang et al. [15]).

The upper reaches of the Laoyunliang River are fed by the outflow of the Fangjiayiu Reservoir, which in turn is fed by rainwater from the neighbouring mountains. In the Xiaopuji district, waste water from a copper factory and municipal sewage are added, which is about 95% of the whole discharge to the Caohai (He et al. [12]). Over a length of 11 km, it flows through the Kunming urban area and is polluted with sewage from settlements along the entire flow path. Its catchment area is home to around 250,000 inhabitants (Li et al. [20]). It flows downstream of the municipal wastewater treatment plant No. 3 into the Caohai.

The 3.7 km long Wulong River used to be fed by the wastewater of the Yunda Hospital. According to (Yang [45]), however, the source has been relocated to the Baima District and is only polluted with waste water from the hospital during rainy weather. (Yang [45]) was able to show that the water quality has improved considerably since 2009, and most recently quality level IV according to the Chinese classification system was achieved. It flows into the Dianchi near the Lijia Park.

According to Ding et al. [5] the Daguan River is an artificial river that flows together from various sources in Kunming city centre and is called the Daguan from the Guan Building onwards. He et al. [12] showed that nearly 100% of the water in the Daguan comes from the Niulanjiang River replenishment. The Daguan River is about 3 km long and flows into the Caohai in Daguan Park. The park originated from a wetland area that was in a very poor condition in the past and was restored as part of an ecological upgrading project (Ma et al. [22]).

The Xiba River is a side channel of the Niulanjiang River, which is known for its heavy pollution and brings high nutrient loads from the mountains in the east of the city, which are mainly agricultural. The Niulan River Diversion Project alone has introduced 400 million m^3 of water into Dianchi in 2015.

In the Chuanfang River the WWTP 1 of Kunming accounts for about 95% of the discharge (He et al. [12]). Its water quality is therefore directly dependent on the purification capacity of this WWTP.

Obviously, the tributaries of the Caohai are anthropogenically dominated. Three rivers are mainly dominated by WWTP effluents, two rivers receive huge amounts of water from the Niulanjiang River replenishment (He et al. [12]). Only the Xinyunliang River has a more natural dynamic.

3.2.2 Materials and Methods

3.2.2.1 Sampling

3.2.2.1.1 Water Sampling
Water samples were taken both at the seven tributaries of the Caohai and its outflow, and on the lake. The timing of sampling was based on the hydrographs of rainfall in the city of Kunming (see Fig. 3.3). The water bodies were sampled at the peak of the rainy season and in the two transitional phases before and after the rainy season.

At the tributaries, the sampling points were selected so that they were as close as possible to the mouth of the Caohai in order to record all pressures on the lake. On the other hand, an influence of wind-induced backwash of water from the lake into the tributaries had to be avoided. The third constraint for the selection of the sampling points was the accessibility of the tributaries. Accordingly, the sampling points marked R in Fig. 3.2 were selected. These restrictions also applied to the outflow of the Caohai.

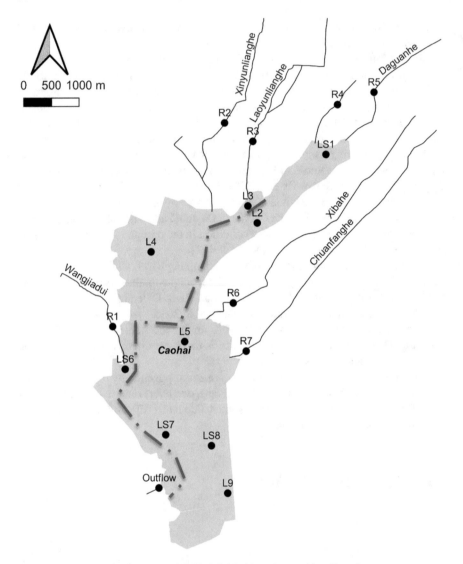

Fig. 3.2 Western and eastern part of Caohai divided by a dam and its tributaries

For the sampling on the Caohai, the premise was to ensure an even distribution of the samples. The scheme was deviated from the points where the planned dredgings took place, because sediment samples were also taken there. Furthermore, the aforementioned division of the Caohai had to be considered. The distribution of the sampling points can be seen in Fig. 3.2.

The water samples were taken in a depth-integrated manner with an immersion bottle, between the surface and the bottom of the water, as described in DIN 38402 (1985). The samples were stored in PE bottles. Together with the sampling, the

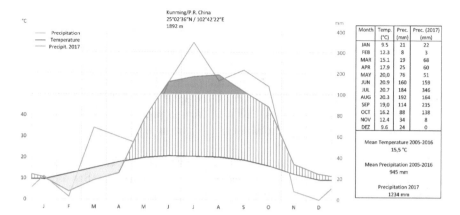

Fig. 3.3 Kunming hydrograph

accompanying parameters oxygen (optical), pH value, conductivity, redox potential (all three electrochemical) were analysed by using probes. These parameters serve to interpret and check the plausibility of the measurement results.

In the laboratory of the Kunming Institute of Environmental Science (KMIES) the samples were prepared for further analysis. For this purpose each sample was divided into two subsamples. Subsample 1 was used unfiltrated for the analysis of the parameters total nitrogen and total phosphorus. Subsample 2 was filtrated through a 0.45 μm filter and served as analysis sample for the parameters ammonium, nitrite, nitrate, orthophosphate and total dissolved phosphorus.

3.2.2.1.2 Sediment Sampling

In order to assess the nutrient release of the sediments of the Caohai, their sampling was carried out in parallel with the water sampling described in chapter Water Sampling. The sediment sampling took place in May and July 2017. Sediment samples were taken at four different locations. The local authorities planned dredging at various locations in the Caohai at the time of sampling. The sampling sites were selected according to these planned dredging operations (see Table 3.1, Fig. 3.2).

The sampling was carried out from a boat parallel to the sampling of the water. A Van-Veen bottom grab was used for this purpose. The samples were individually packed in plastic bags and analysed in the laboratory of the Agro-Environmental Protection Institute in Dali.

Table 3.1 Sediment dredging sites at Caohai in the year 2017

Sample	Dredging	Coordinates
LS1	Projected	25°1.323 N, 102°40.058 E
LS6	Not projected	24°59.358 N, 102° 37.866 E
LS7	Projected	24°58.731 N, 102° 38.279 E
LS8	Not projected	24°58.616 N, 102°38.762 E

3.2.2.2 Analytics

3.2.2.2.1 Water Analytics
See Table 3.3

For the water analytics we used a spectral photometer using methods equal to international, European or German standards (compare Table 3.2).

For the most parameters the reagents were used in form of powder pillows. For the parameters total nitrogen and total phosphorus we used vial tests. These samples were digested using a vial cooker.

We measured each sample three times for statistical confidence. Presented results are arithmetic mean values. Organic nitrogen was calculated as the difference between total nitrogen and its ionic forms.

3.2.2.2.2 Sediment Analytics
To measure the sediment nutrient release we used acrylic glass tubes with a diameter of 6.7 cm and a volume of one litre. Each sample was divided into two subsamples for aerobic and anaerobic tubes. The samples were carefully covered with deionised water. Additionally we prepared two blind values. We used an air pump to ventilate aerobic samples during the measurement period. There was daily measurement of oxygen concentration, redox potential and temperature in both sampling rows. After 3, 7 and 11 days we analysed ammonium, nitrite, nitrate, orthophosphate, total dissolved phosphorus and total phosphorus like presented in Chap. 3.2.2.1.

Sediment nutrient release was extrapolated to the surface of the Caohai using the following formula:

$$Sediment\ nutrient\ release\left[\frac{mg}{m^2*d}\right]=\frac{(Concentration_0-Concentration_1)\left[\frac{mg}{l}\right]*Volume\ [l]}{Surface\ Area\left[m^2\right]*Experiment\ time[d]}$$

Table 3.2 Parameters and standards used to conduct spectral photometry for water samples

Group	Parameter	Standard
Nitrogen	Ammonium	DIN 38406-5:1983
	Nitrite	DIN EN 26777:1993
	Nitrate	DIN 38405-9:2011
	Total nitrogen	DIN EN ISO 11905-1:1998
Phoshphorus	Orthophosphate	DIN EN ISO 6878:2004
	Total dissolved Phosphorus	DIN EN ISO 6878:2004
	Total phosphorus	DIN EN ISO 6878:2004

Table 3.3 Water analytics results

Sp. Point	Month	pH	EC	Redox	O2	Temp	NH4-N	NO2-N	NO3-N	TN	PO4-P	TDP	TP
Wangjiadui	May	7.8	742	nm	3.9	18.6	7.0	nm	0.4	19.9	0.1	0.1	0.7
	July	7.7	881	116	4.0	21.6	8.4	0.02	0.3	11.2	0.6	0.7	0.9
Xinyunliang	May	7.7	634	143	0.3	20.7	2.4	0.08	0.8	7.7	0.3	0.3	1.4
	July	7.7	604	-6	3.3	20.0	1.8	0.09	0.7	5.7	0.3	0.3	0.4
	September	7.8	481	118	3.7	20.1	2.8	0.15	1.8	7.0	0.0	0.2	1.3
Laoyunliang	May	7.3	542	81	3.7	21.5	4.8	0.08	2.4	16.0	0.2	0.2	0.9
	July	7.4	702	109	3.2	21.3	4.1	0.20	3.2	16.5	0.3	0.4	0.6
	September	7.5	571	141	5.6	21.4	1.5	0.17	2.0	9.2	0.6	r	1.1
Wulong	May	7.3	560	131	7.5	21.5	2.1	0.04	3.8	14.0	0.1	0.1	1.0
	July	7.2	696	188	5.4	21.4	0.4	0.13	3.2	14.7	0.3	r	0.3
	September	7.2	620	119	5.5	21.9	1.7	0.37	4.6	14.0	0.1	r	0.6
Daguan	May	7.9	382	111	4.5	17.3	0.9	0.02	1.7	5.6	0.1	0.1	0.5
	July	7.5	435	175	0.5	21.4	1.4	0.16	1.5	9.8	0.7	r	r
	September	7.8	303	36	1.8	20.5	0.4	0.10	1.0	5.2	3.4	r	r
Xiba	May	8.7	405	102	11.3	16.7	0.1	0.08	1.3	3.4	0.1	0.1	0.3
	July	8.4	733	135	16.5	24.0	0.1	0.03	1.2	3.0	0.1	r	r
	September	8.1	362	158	4.5	21.5	0.9	0.24	2.0	3.7	0.3	r	0.8
Chuanfang	May	7.8	524	148	5.7	18.7	0.5	0.05	1.7	6.1	0.0	0.0	0.3
	July	7.6	661	135	6.3	24.4	nm	0.07	1.8	7.5	0.1	0.1	0.2
	September	7.8	475	132	3.8	21.5	0.2	0.11	1.6	4.6	r	0.3	0.3
Caohai 1	May	8.0	391	104	5.0	16.8	0.7	nm	1.8	3.5	0.1	0.1	0.3
	July	7.6	302	133	1.7	19.8	2.1	0.09	0.8	5.8	0.3	r	0.4
	September	7.8	358	70	2.9	26.5	0.6	0.06	1.4	4.3	0.0	0.6	r
Caohai 2	May	8.4	427	105	5.9	18.6	0.1	0.08	1.8	3.8	0.0	0.1	0.6
	July	8.6	394	112	9.9	22.6	0.1	0.08	1.4	8.7	0.1	r	0.4
	September	8.4	362	157	8.5	23.5	0.1	0.08	1.0	3.8	0.1	0.3	0.3
Caohai 3	May	7.6	682	36	3.1	17.7	4.5	0.79	3.6	34.4	0.5	0.7	r
	July	7.4	444	142	1.5	21.8	3.0	0.14	1.7	13.2	0.3	r	0.4
Caohai 4	May	8.7	541	36	5.4	19.7	0.1	0.10	1.4	4.8	0.1	0.2	0.8
	July	7.9	501	56	4.4	21.4	1.5	0.37	1.7	8.3	0.3	r	0.2
Caohai 5	May	8.4	349	57	8.8	20.1	0.4	0.04	0.8	1.6	0.1	0.2	0.3
	July	9.0	393	66	10.7	23.6	0.1	0.10	0.6	9.2	0.1	0.3	0.5
	September	8.7	360	145	9.4	25.0	0.2	0.10	1.4	4.4	0.1	0.2	0.6
Caohai 6	May	8.7	531	63	6.2	19.4	0.3	0.09	1.3	4.9	0.2	r	r
	July	8.2	543	139	10.0	22.7	0.2	0.26	1.7	10.5	0.1	r	0.4
	September	8.2	555	168	3.6	24.0	1.0	0.36	2.5	15.4	0.0	0.3	1.6
Caohai 7	May	8.9	301	-28	12.6	19.9	0.1	0.04	0.6	3.1	0.1	0.1	0.4
	July	9.0	383	113	9.8	24.3	0.5	0.11	1.1	4.8	0.1	r	0.2
	September	8.7	354	106	7.0	23.5	0.3	0.09	1.1	3.5	0.0	0.2	1.3
Caohai 8	May	8.9	333	9	8.9	20.4	nm	0.04	1.1	2.8	0.1	0.1	0.3
	July	8.4	425	167	6.8	23.9	1.3	0.16	0.7	4.7	0.1	r	0.3
	September	8.3	371	113	6.2	nm	0.3	0.11	1.3	4.2	0.1	r	0.4
Caohai 9	May	9.4	289	70	12.2	20.2	0.1	0.02	0.7	3.4	0.1	0.1	0.4
	July	8.0	404	170	3.9	23.0	0.7	0.13	0.5	5.0	0.1	r	0.3
	September	7.9	386	104	3.2	23.2	0.1	0.16	1.7	4.0	0.1	0.1	0.4
Outflow	May	9.5	339	82	16.5	21.2	0.1	0.05	0.8	7.8	0.1	0.1	0.3
	July	8.0	505	118	6.8	22.1	3.0	0.20	1.4	5.7	0.1	0.3	0.4
	September	7.8	549	131	4.6	22.9	0.2	0.39	2.4	8.0	0.3	r	0.5

pH-value dimensionless, electric conductivity in µS/cm; redox potential in mV, Temperature in °C, all other values in mg/l

nm = not measurable; r = rejected

Table 3.4 Environmental quality standards for surface water (Chinese standard: UDC614.7 (083.75) GB 3838-88)

Parameter	Unit	Class I	Class II	Class III	Class IV	Class V
COD	$\leq$mg/l	15	15	20	30	40
BOD$_5$	$\leq$mg/l	3	3	4	6	10
Diss O$_2$	$\geq$mg/l	7.5	6	5	3	2
TN	$\leq$mg/l	0.2	0.5	1	1.5	2
NH4-N	$\leq$mg/l	0.15	0.5	1	1.5	2
TP	$\leq$mg/l	0.02 (lakes, dams 0.01)	0.1 (lakes, dams 0.025)	0,2 (lakes, dams 0.05)	0.3 (lakes, dams 0.1)	0.4 (lakes, dams 0.2)

Table 3.5 Compares between aerobic and anoxic milieu

		NH4-N	NO2-N	NO3-N	PO4-P	TDP	TP
Aerobic							
	1 May	130,3	3,79	14,6	31,9	33,7	24,9
	July	26,8	0,06	46,7	11,4	-2,0	-3,4
	6 May	78,3	0,17	12,8	27,8	45,1	34,7
	July	-3,3	-4,36	-4,6	19,2	21,0	10,0
	7 May	91,4	0,25	1,3	21,5	38,2	33,1
	July	22,7	-0,76	11,9	11,0	6,8	11,7
	8 May	30,8	1,81	-17,5	15,9	-8,1	-5,1
	July	8,1	-0,77	3,5	14,2	-1,0	0,2
Anoxic							
	1 May	102,8	0,17	13,6	20,2	16,0	17,3
	July	-1,5	38,00	115,3	27,8	18,1	10,2
	6 May	74,2	0,09	6,4	26,8	12,8	17,1
	July	-27,8	-8,59	60,3	22,2	30,5	11,6
	7 May	104,2	-0,17	-2,3	17,3	14,0	13,1
	July	5,1	24,01	77,1	6,6	6,8	6,2
	8 May	54,1	0,04	-1,8	18,8	6,3	9,2
	July	9,4	26,75	76,3	15,3	6,8	7,0

Release Rate 27.07. to 04.08.2017 [mg/m²·d]

3.2.3   Results and Discussion

3.2.3.1   Analytics

3.2.3.1.1 Water Analytics

Lake Caohai forms with its tributaries and its two separated parts three comparable water systems. Following, we will summarize the results of the water analytics, as they are the basis for the shallow lake modelling. In general, there are few sources in the literature available for both the Caohai and its tributaries. Furthermore, the Chinese water quality classification system is important for Chinese decision makers (Su et al. [31]). It uses chemical and biological oxygen demand, dissolved oxygen, ammonium-nitrogen, total nitrogen and total phosphorus concentrations as indicators for classifying water bodies into six classes. "Class I" is the best and "Class V Inferior" the worst (see Table 3.5) (Table 3.4).

In the tributaries the pH-value was 7.7 ($\pm$0.4), in the eastern part of Lake Caohai 8.5 ($\pm$0.47), in the western part 8.1 ($\pm$0.5) and 8.4 ($\pm$0.7) in the outflow. We measured remarkable individual values in the Xiba River of 8.7 in May and in the Lake Caohai of 9.4 in May. These results shows the huge impact of algae in Lake Caohai shifting the pH to acidic conditions by uptake of CO_2. The electric conductivity and the redox potential were unremarkable in most samplings. In the tributaries we measured the dissolved oxygen concentration to 5.1 ($\pm$3.5) mg/l, to 7.4 ($\pm$3.1) for the eastern part of Lake Caohai, to 4.9 ($\pm$2.5) for the western part of Lake Coahai. Especially in the highly polluted western part the oxygen concentration was lower compared to the other parts. The highly turbulent outflow had a concentration of 9.3 ($\pm$5.2).

The ammonium concentration in the tributaries was 2.2 ($\pm$2.3) mg/l. It became obvious that the rivers Wangjiadui, Xinyunliang and Laoyunliang, which are flowing in the western separated part of Lake Caohai, had much higher ammonium concentrations. For this group the mean was 4.1 ($\pm$2.35) mg/l. For the second group, containing the rivers Wulong, Daguan, Xiba and Chuanfang the mean value would be 0.8 ($\pm$0.65) mg/l. This difference was calculated to be highly statistically significant (p < 0.001). Nitrite concentration in the tributaries was 0.11 ($\pm$0.08) mg/l, 0.09 ($\pm$0.04) mg/l and nitrate concentration was 1.9 ($\pm$1.1) mg/l. At the Wulong river we measured a remarkable concentration of 4.6 mg/l in September. The concentration of total nitrogen was 9.2 ($\pm$4.9) mg/l in the tributaries. Again, the rivers Wangjiadui, Xinyunliang, Laoyunliang, and also, Wulong formed a group with higher mean values of 12.3 ($\pm$4.3) mg/l. Unlike, the group of other rivers had a mean value of 5.4 ($\pm$2.05) mg/l, being a statistical highly significant difference (p < 0.001).

Orthophosphate was 0.4 ($\pm$0.72) mg/l in the tributaries, 0.1 ($\pm$0.1). The highest value measured was 3.36 mg/l in the Daguan River in September. Eliminating this one value as an outlier in the river dataset lowers the mean concentration to 0.2 ($\pm$0.19) mg/l. Total phosphorus was 0.7 ($\pm$0.4) mg/l in the tributaries. Yang and Jin [44] made investigations between 2005 and 2007 at the Caohai. For nearly all tributaries, the TN values dropped in this period. An exception was the Wangjiadui River, which in this study had a mean TN concentration of 9.5 mg/l over three years. Compared to our results there was no improvement of the water quality in the meantime. The water quality of the Wangjadui River was Class IV regarding oxygen and water quality class V inferior in all nutrient parameters.

The Xinyunliang River was described at Huang et al. [14] to be the most polluted tributary of Lake Caohai. For 2012 they reported ammonium to be 6.92 mg/l, TN to be 13.40 mg/l and TP to be 0.38 mg/l. For the same year Nie [24] reported values for the rainy season where ammonium was 11.53 mg/l, TN 12.01 mg/l and TP 0.90 mg/l. Based on our results, we can assume that the actions that had been made to restore the river showed some success. Ammonium concentration dropped to only 20% of the former concentration reported. TN dropped to 56% of the values reported. There was no change in the concentration of TP. In general, the water quality became better but it is still poor. The Xinyunliang River was class V inferior in all aspects.

Zhiyi et al. [49] performed some measurements at the Laoyunliang River. They reported that ammonium was 1.55 mg/l, TN was 10.15 mg/l and TP was 0.26 mg/l as mean values from three measurements. Compared to this in our results ammonium and TP were significant higher. TN dropped 33%, which is a better water quality than in 2016, but still alarmingly high. The Laoyunliang River was class IV in the oxygen and class V inferior in all nutrient parameters. For the other parameters it was class V inferior having the highest nutrient concentration of TN and TP of the tributaries of the western part of Lake Caohai.

Yang et al. [45] gave a good overview over the development of the water quality of the Wulong River between 2005 and 2015. Thus, ammonium had a peak in 2008 with 33 mg/l and TP was 3.6 mg/l. Until 2015 the concentration dropped to 1 mg/l for ammonium and 0.25 mg/l for TP. In our measurements ammonium was worse with 1.37 mg/l and TP 0.62. Especially TN concentration, which are not reported yet in literature, showed high concentrations of of 14.21 mg/l. Regarding this value, the Wulong river was one of the most polluted rivers in our study. The Wulong river reached water quality class III regarding the oxygen concentration.

There are few references regarding the Daguan River. There are few references regarding the Daguan [22] measured the water quality in February and March 2010 when the Wujiadui theme park was built. For the inflowing water from the Daguan River Ma et al. [22] measured ammonia concentrations of 6.26 mg/l and TP concentration of 5.65 mg/l. Compared to these values the water quality was much better in our study with ammonia being 0.90 mg/l and TP being 0.45 mg/l. Compared to some unpublished results from our scientific partners from the year 2015, we can assume that our values are plausible and that the Daguan Rivers water quality became better since 2010. For ammonia in our study, the river had a medium condition in May and July and good condition in September being one of the rivers with a better water quality. Nevertheless, the Daguan River is still one of the biggest pollution sources for Lake Caohai because of its huge discharge. The Daguan River had class V inferior for all parameters except of ammonium, where it had class IV and TP with class V.

The Xiba River had class IV in general for the oxygen, class III for ammonium and class V inferior for the other parameters. Looking to May and July it had water quality class I for oxygen and ammonium, but because of the high pH-Values and the ammonia-ammonium-equilibrium this class cannot be given.

Zhan et al. [48] reported an improved water quality for the Chuanfang River after a river rehabilitation. Compared to those 2014 values in our 2017 sampling campaign TN dropped around half. TP has not dropped and is still the same, as the Chuanfang River is still the outlet of the wastewater treatment plant number 1. The Chuanfang River had class IV in oxygen, II in ammonia, V in TP and V inferior in the other parameters. These four rivers flow to the western part of Lake Caohai. In this part of the lake the water quality class was V inferior for TN and TP. In the south east of the Caohai the water quality class was better regarding oxygen (between I and III) and ammonia (between I and IV). This picture could also be seen in the water quality of the outflow, which was in general class V inferior for the most parameters, but better for oxygen (I to IV).

In the eastern part of Lake Caohai the ammonium concentration was 0.4 (±0.5) mg/l, 1.5 (±1.5) mg/l in the western part and 1.1 (±1.3) mg/l in the outflow. Nitrite was 0.09 (±0.04) in the eastern part of Lake Caohai, 0.30 (±0.23) mg/l in the western part of Lake Caohai and 0.21 (±0.14) mg/l in the outflow. The nitrate concentration was 1.1 (±0.4) mg/l in the eastern part of Lake Caohai, 2.0 (±0.7) mg/l in the western part of Lake Caohai and 1.5 (±0.7) mg/l in the outflow. The concentration of total nitrogen was 4.5 (±1.8) mg/l in the eastern part of Lake Caohai, 13.1 (± 9.4) for the western part and 7.2 (±1.1) for the outflow. For Orthophosphate the concentration was 0.1 (±0.1) in the eastern part of the lake, 0.2 (±0.2) mg/l in the western part and 0.1 (±0.1) mg/l in the outflow.

For the lake Caohai Jia (2018) reported measurement results for total nitrogen and total phosphorus from 1993 to 2017. The water quality became better since the year 2011, when TN dropped to 6.1 mg/l and TP to 0.22 mg/l. The reported value for 2017 was 3.7 mg/l for TN and 0.15 mg/l for TP. Our result from the same year for the Caohai were 4.5 mg/l for TN and 0.44 mg/l for TP in the eastern part and 13.6 mg/l for TN and 0.9 mg/l for TP in the western part. As the reported values in the literature were yearly averaged values one can see that in the rainy season nitrogen and phosphorus values were significant higher in the rainy season, especially in the western part of Lake Caohai. We could show that nutrient intake from the rivers is still very high.

For the western part of Lake Caohai at the three sampling points, the water quality was different for oxygen ranging from class V inferior in the north to class III in the south. Nevertheless, the nutrient parameters were still class V inferior. A decrease in ammonium concentrations over the flow path could be seen. From the north (class V inferior) to the south (class III) ammonium concentrations dropped. Our results show not only that for a lot of parameters during the rainy season the water quality is in the worst class, but also that the difference to the threshold of class V is still remarkable. For example, TN in the middle of the western part of the Caohai was 34.4 mg/l whereas the threshold value is 2 mg/l. For TP the highest value was 1.58 mg/l in the same part of the lake and the threshold is 0.2 mg/l for water quality class V. The objective of the thirteenth five-year plan is to achieve the water quality class V until 2018 for the Caohai. Studying our data, which we collected as stich samples in the rainy season, one can see that there is a need for further effort to come closer to this objective. Our new data acquisition together with other available data now is the main input for the Caohai modelling described and discussed below.

3.2.3.1.2 Sediment Analytics

We measured dissolved oxygen, redox potential and temperature each day over the experiment's period as control parameters for the milieu as we had an aerobic sample row and an anoxic sample row. The full results can be seen in Table 3.1 of the Annex. In the aerobic row the dissolved oxygen concentration was always between 6.9 and 7.3 mg/l. As Kunming lies on a height of around 1,890 m above sea this equals a saturation between 98.5 and 104.2%. When the process became stable from day 2 on, the redox potential ranged between 89 and 115 mV. Because of these facts, we assume that the aerobic milieu was stable in the whole period. The temperature ranged in small interval between 21.4 and 22.3 °C.

In the anoxic sampling row, the dissolved oxygen was under 0.7 mg/l at day 1 and dropped in most samples to 0.2 mg/l at day 2. In one sample, it dropped to 0.3 mg/l. In three samples, the dissolved oxygen dropped slower. Sample by sample it dropped under 0.2 mg/l. The last sample did not reach this value before day 9. The redox potential ranged from day 2 between -124 and -63. At the last day of our experiment, we measured a redox potential of -423 in one sample, which we assume as an outlier. In most cases, the anoxic milieu was stable aside from those samples the dissolved oxygen dropped slower. Similarly, like in the aerobic sampling row, the temperature ranged between 21.4 and 22.3 °C.

As we took the samples in two different months, we tested first if there were significant differences between the two monthly datasets. If so, we report the results separated by month. For the aerobic samples the difference between the two months was significant ($p < 0.05$) for ammonium and nitrite, but not for nitrate ($p > 0.05$). Ammonium sediment release was 82.7 (± 35.6) mg/m^{2}d for the samples taken in May and 13.6 (± 11.9) mg/m^{2}d for the samples taken in July. For nitrite, it was 1.51 (± 1.47) mg/m^{2}d for May and -1.46 (± 1.71) mg/m^{2}d for July. For nitrate, the mean value for all samples was 8.59 (± 17.5) mg/m^{2}d.

For the anoxic sampling row, the difference for ammonium between the May and July set was very significant ($p < 0.01$), for nitrite it was not significant ($p > 0.05$) and for nitrate it was very significant again ($p < 0.005$). For ammonium, the sediment release rate was 83.8 (± 20.09) mg/m^{2}d for the May data and -3.7 (± 14.5) mg/m^{2}d for July. Nitrite was 10.0 (± 15.8) mg/m^{2}d for the whole data set. Nitrate was 4.0 (± 6.6) mg/m^{2}d for the May dataset and 82.3 (± 20.2) mg/m^{2}d for the July dataset.

For ammonium and nitrite there was no significant difference in the sediment release rate between the aerobic and the anoxic samples ($p > 0.1$). For nitrite it was very significant ($p < 0.005$).

For nearly all three parameters in both milieus and in both months the sediment release rate in the northeast of the Caohai, where the Daguan river flows into the Lake (sample 1) was much higher than nearby the outflow (sample 8). An exception is ammonium in the anoxic milieu in July.

Looking at the phosphorus parameters there were nearly no significant differences between the results of May and July. The only exception was total phosphorus in the anoxic milieu ($p < 0.05$). The sediment release of orthophosphate in the aerobic milieu was 19.1 (± 7.1), 16.72 (± 19.2) mg/m^{2}d for the total dissolved phosphorus and 13.24 (± 15.0) mg/m^{2}d for total phosphorus. In the anoxic milieu, orthophosphate was 19.4 (± 6.3) mg/m^{2}d and total dissolved phosphorus was 13.9 (± 7.6) mg/m^{2}d. Total phosphorus was 14.2 (± 3.3) mg/m^{2}d in May and 8.8 (± 2.2) mg/m^{2}d in July.

For all three parameters there was no statistical significant difference between the aerobic and the anoxic milieu ($p > 0.5$). In the aerobic May dataset and both anaerobic datasets, the sediment release rate was higher at the Daguan inflow in the northeast of the Caohai than at the outflow.

Based on our results we have strong indications for the assumption that the sediment has a huge impact on the water quality of the Lake Caohai. As there were enormous nutrient sediment releases measured in the anoxic, as well as in the aerobic, samples, we strongly recommend to take an even stronger focus on the sediment in future lake management. As it is not the expected behavior of the sediment to show such high sediment release rates in the aerobic milieu, we also strongly encourage other scientist to repeat and further test the sediment of Dianchi to gain a more detailed picture and to scrutinize our results.

3.3 Study 2: Shallow Lake Modelling of Lake Caohai

As already mentioned, the water quality of Lake Caohai and its tributaries was proven to be to be strongly eutrophicated. The heavy and complex burden of water pollution do not make simple answers very likely. From other examples like Lake Constance (Bloesch and Schroeder [1]), one can learn that the basis for a successful water management lies in strong management efforts. They are consisting of thorough and ongoing investigations to establish the best system understanding, consistent management decisions and long-term commitment for decades.

Thus, in the frame of the Sino-German project SINOWATER Chinese and German partners together discussed several conceivable management scenarios. To evaluate the effects of the different management scenarios we decided to use a dynamic simulation approach.

Numerical models are very useful management tools in lake management, as they aggregate existing knowledge and hypotheses and thus allow the dynamic simulation of different management scenarios. For the SINOWATER project, we used the Stoichiometric Lake Model (StoLaM, Strauss [29] and Strauss et al. [30]), which is designed to deal with deep as well as shallow lakes is the context of sustainable lake management. As the basis for the dynamic lake modelling, StoLaM includes the hydrodynamic sub-model HyLaM, which is able to calculate the hydrodynamic conditions with high temporal and spatial resolution even in shallow lakes. StoLaM describes several nutrient cycles quantitatively taking into account the sediment–water interactions and the role of plankton and omnivorous fish in the food web.

The Dianchi was several times modelled with different lake models (Vinçon-Leite and Casenave [34]). However, especially for a system in which phosphorus and nitrogen in combination determine the growth of algae, we consider the use of models with phytoplankton modules as reasonable, which integrate dynamic nutrient storage as well as stoichiometric principles. Although these processes have been known and considered as relevant since a longer time, they are rarely integrated even in complex ecosystem models (Jørgensen [17] and Strauss et al. [30]). We see the use of StoLaM, which combines such features with a hydrodynamic 1D modelling approach, as a relevant contribution to further improving management decisions for Lake Dianchi.

The setting up of the StoLaM for Lake Caohai was a two-step approach:

In 2016, the model was parameterised with the available hydrological and climatic basic information and starts with default estimations on the nutrient cycles, sediment interactions and plankton 2016. Based on the first draft model we presented first findings in the 2016 SINOWATER Dianchi Shallow Lake Symposium at Kunming. At this time the values for the lake-internal sediment nitrogen and phosphorus release rates were as unknown as the nutrient loss due to sedimentation of detritus-bound nutrients from the inflows. Furthermore, a detailed picture of the composition of those nutrients in the inflowing waters was not available and we could not estimate the denitrification rate due to this lack of data.

Through the SINOWATER sampling campaign in 2017, a broader data set could be used for the parameterization and validation of the entire model, which was adapted to the specific simulation tasks in this project. This data included the new water quality data from the Kunming Institute of Environmental Sciences as well as our own monitoring data as described above, with which it was possible to calculate management scenarios more properly.

Together with our Chinese project partners we have developed scenarios that are based on previously discussed water management strategies in Kunming. Finally, five main scenarios were selected for numerical modelling, with some of the main scenarios being divided into sub-scenarios as described below:

1. **Standard scenario**,
   modelling the status quo in a high temporal (at least hourly time steps) resolution
2. **Sediment nutrient release Scenario**,
   is dealing with changes in the nutrient release of the sediment as they could occur after sediment removal by dredging measures
3. **Tributary Water Quality Scenario**,
   is dealing with changes in the nutrient inflow through the tributaries as they could occur after improvements in the sewer system or upgrading waste water treatment plant
4. **Tributary Water Quantity Scenario**,
   is dealing with changes in the amount of water flowing into Lake Caohai through the tributaries as they could occur for several natural or anthropogenic changes in the catchment controlling the wash out of nutrients
5. **Caohai Water Level Scenario**,
   is dealing with changes in the water level and thus the water volume in the lake

First of all, we want to introduce the conceptual structure and theoretical foundations of the model StoLaM. Secondly, we show the data used and the results obtained. Finally, we discuss the results and management scenarios.

3.3.1 Theoretical Foundation and Adaptions

The Stoichiometric Lake Model (StoLaM, Strauss [29] and Strauss et al. [30]) is a general 1D hydrodynamic-ecological water quality model for standing waters. It includes the hydrodynamic lake model HyLaM (Strauss [29], validated by Gross [8]), which generates the underwater light regime, water temperature and turbulence in the vertical water column using measured weather data as input (compare Fig. 3.4). With this approach, it is possible to simulate a realistic physical environment for the plankton organisms with a high temporal resolution. Based on these hydrodynamic processes, the modelling concept takes into consideration the zooplankton and phytoplankton succession, the quantitative cycles of several nutrients (P, N, Si), the oxygen concentration, the sediment–water interaction and the role of omnivorous fish in food web interactions. StoLaM integrates the principles of ecological stoichiometry, which means the changing ratios of nutrients such as carbon, phosphorus, nitrogen and silicon within the food web and their impact on the growth rates of plankton organisms with respect to nutrient limitation. Beneath the weather conditions and the ingestion by herbivorous zooplankton, the phytoplankton growth is based essentially on the amount of phosphorus, nitrogen and silicon, which are potentially limiting nutrients. The external nutrient concentration and the variable internal nutrient pool (cell quota) of the algae controlled the uptake rates of phosphorus and nitrogen from the water phase and storage by the algae. Thus, the growth rate of the algae depends on the internal nutrient cell quota. In this approach, the carrying capacity of the algae depends on the amount of the most limiting nutrient stored in the algal biomass (for further particulars of the complete model, see Strauss [29] and Strauss et al. [30]).

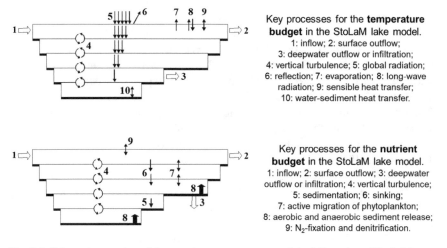

Key processes for the **temperature budget** in the StoLaM lake model.
1: inflow; 2: surface outflow;
3: deepwater outflow or infiltration;
4: vertical turbulence; 5: global radiation;
6: reflection; 7: evaporation; 8: long-wave radiation; 9: sensible heat transfer;
10: water-sediment heat transfer.

Key processes for the **nutrient budget** in the StoLaM lake model.
1: inflow; 2: surface outflow; 3: deepwater outflow or infiltration; 4: vertical turbulence;
5: sedimentation; 6: sinking;
7: active migration of phytoplankton;
8: aerobic and anaerobic sediment release;
9: N_2-fixation and denitrification.

Fig. 3.4 Schematic overview of the most important processes of the full version of StoLaM

Due to the modular character of the model, the complexity of the modules used can be easily adapted to the problem definition and data availability. In this study, a simplified version of the interactions and food web structure in StoLaM was used. The following assumptions reduced the complexity of the standard model:

In the Caohai, the blue-green algae (cyanobacteria) *Microcystis aeruginosa* is the predominant species, which also caused the main water quality problems.

Since data on the exact composition and interaction of other organisms were lacking, the implemented complex food web of StoLaM was massively reduced in this model application. For the phytoplankton community, we only considered the population dynamics of *Microcystis aeruginosa*, and no other organism such as zooplankton or fish has been modelled. The dynamics of the cyanobacteria *Microcystis* can also be described without further food web interactions such as zooplankton grazing. This procedure is also justified by the fact that larger colonies of the cyanobacterium *Microcystis* are not edible for zooplankton due to their size, and in some cases also due to excreted toxins (e.g. microcystin). Thus, the chemical-physical environmental conditions and the residence time of the water body are particularly most relevant for modelling the *Microcystis* dominance in Lake Caohai. The silicon dynamics, which are only relevant for diatoms, was also switched off. The physico-chemical processes in the model were otherwise left unchanged. For a better comparison with measured field data, the biomass of *Microcystis* was converted and displayed as chlorophyll-a concentration.

3.3.2 Data input and model parameterisation

In general, we simulated the eastern free flowing part of the Caohai. We included the rivers Wulong, Daguan, Xiba and Chuanfang as inflows.

A simple, box-shaped lake morphometry with 5 vertical layers of 50 cm height each and a uniform surface of 6,790,000 m^2 was used for this model application. For the simulation of the water temperature with the HyLaM model we used meteorological data of the year 2016 provided by the Kunming centre weather station (air temperature [°C], wind speed [m s^{-1}], atmospheric pressure [hPa], and air humidity [%], time resolution of the measurements: 3 h). The global radiation [J cm^{-2} s^{-1}] was calculated on an hourly basis using the formulas of Ryan & Stolzenbach [28], taking into account a typical monthly averaged cloud cover (see Table 3.6). The water attenuation coefficient was set to 0.3 m^{-1}.

To calculate mass flow rates from the concentrations we used river discharges provided by our Chinese partners from the Kunming Institute of Environmental Sciences (compare Fig. 3.5). The specific dynamics of these rivers are reported in a recent paper of these partners (He et al. [12]). An essential input for the StoLaM model is the nutrient data. As it was needed at least as monthly values we took results from the monthly measurements of the Kunming Institute of Environmental Sciences. The proportion of detritus-bound phosphorus (65% of the total phosphorus) and nitrogen (76% of the total nitrogen) was calculated by subtracting the total fractions and the

Table 3.6 Long-term values of mean monthly cloud cover between 2002 and 2015 from the NASA data archive for the Dian Lake Basin region.

Month of the year	Mean cloud cover [%]
1, 2	30
3, 4, 12	45
5	60
6, 7, 10, 11	75
8, 9	85

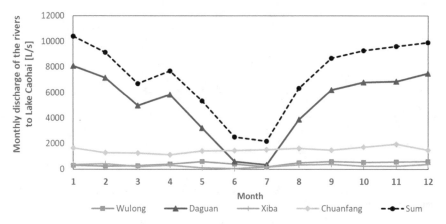

Fig. 3.5 Discharge hydrographs in 2016 used for the simulation provided from the Kunming Institute for Environmental Sciences

dissolved inorganic fractions (compare Table 3.7). The knowledge of the detritus-bound fraction of the inflowing nutrients is extremely important for the simulation results, because a relatively fast sedimentation can be assumed for detritus, whereas the dissolved fractions are available to the algae as nutrients in the water column.

The daily remineralisation rate of detritus-bound nutrients in the lake at 20 °C was calibrated within this project and was set to 20% for carbon, 25% for nitrogen and 20% for phosphorus (temperature coefficient of 1.13, Strauss et al. [30]. The standard detritus-settling rate was assumed as $1 \, \text{m day}^{-1}$. In our modelling approach, there was no additional nutrient release assumed by remineralisation of resuspended sediment in the water column.

For the sediment nutrient release rates, we used our measurements in 2017. As described above, experimental data on sediment release was inhomogeneous. Thus, we took the arithmetic mean values to get robust parameters for the sediment release: For nitrogen the sediment release was 83 mg NH_4-N $m^{-2}d^{-1}$ and for phosphorus it was 19 mg PO_4-P $m^{-2}d^{-1}$. We used these measurements as baseline values at 20 °C in the model, and simulated the temperature-dependent release with an implemented temperature dependency (temperature coefficient of 2.42, Strauss et al. [30]). In the release experiments, we did not found significant differences between aerobic

Table 3.7 Mean nutrient concentrations of the combined external Inflows per year and month (measured in 2017).

	mg PO$_4$/L	mg Detr-P/L	mg NO$_3$/L	mg NH$_4$/L	mg Detr-N/L
Annual mean	**0.066**	**0.212**	**1.65**	**0.76**	**4.53**
January	0.063	0.197	1.52	0.80	4.41
February	0.065	0.196	1.51	0.80	4.38
March	0.064	0.201	1.56	0.79	4.47
April	0.067	0.203	1.58	0.82	4.50
May	0.069	0.224	1.76	0.79	4.78
June	0.067	0.249	1.96	0.62	4.81
July	0.061	0.235	1.79	0.48	4.21
August	0.069	0.216	1.68	0.76	4.53
September	0.069	0.209	1.63	0.81	4.56
October	0.065	0.205	1.60	0.80	4.55
November	0.065	0.206	1.61	0.79	4.55
December	0.067	0.205	1.59	0.82	4.55

and anaerobic conditions. Hence, we used the same release rates for all oxygen concentrations in the model.

The use of measured sediment nutrient release rates in the model, however, is a necessary parameterisation of lake-specific processes. It is important to mention, that the StoLaM model (vers. 3.02.03) was used in this project without any calibration of parameters with one exception: Only the remineralisation rates of detritus-bound nutrients were adjusted to optimize the fit between measured and simulated nutrient concentrations in Lake Caohai. This data gap could be closed by future experiments.

We modelled the non-N$_2$-fixing cyanobacterial species *Microcystis aeruginosa* with a maximum growth rate of 1.5 day^{-1} and a mortality rate of 0.1 day^{-1}. A vertical buoyancy of 1.38 m day^{-1} is used, which leads to an accumulation in the upper water layers at low turbulence of the water body. An overview of all algal parameters used is given in Annex 2.

3.3.3   Model simulation scenarios

As described in the introduction we investigated five main scenarios. The first scenario was the standard run of the simulation with no changes of the parameters described above. We divided the second scenario into one sub-scenario without nutrient release of the sediment (2a) and another sub-scenario with a sediment release of only 75% of the standard to simulate a dredging of 25% of the lakes surface (2b). The third scenario again we divided into two sub-scenarios: One was without external inflow of nutrients (3a), and as the second sub-scenario, we set the nutrient to only 50% of the standard value (3b), both at unchanged water quantities. These scenarios should show the effects of an increased wastewater treatment and decentralized actions in the catchment system. The fourth main scenario we separated into three sub-scenarios. First, we used an annual mean water flow instead of a hydrograph to smooth the variation in discharge (4a). Second, we doubled the inflow from the rivers at constant nutrient concentrations, simulating an increasing number of inhabitants and impervious areas (4b). Finally, we halved the inflow to test the influence of possible sponge city actions, which are discussed in China a lot (4c). The fifth main scenario we separated into two sub-scenarios. We changed the standard water depth from 2.5 to 1.5 m (5a) and to 3.5 m (5b). These options were discussed in Kunming at the time of our project and thus considered.

An important input for the HyLaM module is the theoretical hydrological renewal time of the lake water. It describes the number of days necessary to flush the total volume of the lake water body once. The renewal time was calculated for each scenario in dependence of the inflowing water and the lake volume (compare Table 3.8).

Table 3.8   Water inflow, lake volume and water renewal time

No.	Description of scenario	Sub-scenario	Water inflow [l/s]	Lake volume [Mio. m^3]	Theoretical water renewal time [days]
1	Standard	Satatus Quo nutrient inflow concentration, nutrient release and unsteady inflow; 2.5 m water depth	7,300	16.975	26.9
2a	Sediment nutrient release	No release	7,300	16.975	26.9
2b		75% release (= 25% lake area dredged)	7,300	16.975	26.9
3a	Tributary water quality	0% nutrient inflow	7,300	16.975	26.9
3b		50% nutrient inflow	7,300	16.975	26.9
4a	Tributary water quantity	Continuous average	7,300	16.975	26.9
4b		200% inflow	14,600	16.975	13.5
4c		50% inflow	3,850	16.975	51.0
5a	Caohai water level	1.5 m	7,300	10.815	17.1
5b		3.5 m	7,300	23.765	37.7

Table 3.9 Simulated phosphorus flows and balance (yearly average) in mg P m^{-2} d^{-1}

No	Description of scenario	Sub-scenario	Inflow	Sediment release	Sedimentation	Outflow	Balance
1	Standard		25.3	14.6	23.1	16.4	0.4
2a	Sediment nutrient Release	No release	25.3	–	16.6	8.5	0.2
2b		75% release (= 25% lake area dredged)	25.3	10.9	21.6	14.3	0.3
3a	Tributary water quality	0% nutrient inflow		15.5	7.3	8.4	−0.2
3b		50% nutrient inflow	12.9	14.8	15.6	11.9	0.2
4a	Tributary water quantity	Continuous average	25.8	14.3	23.0	16.8	0.3
4b		200% inflow	51.6	12.3	31.8	31.6	0.5
4c		50% inflow	13.6	15.4	18.5	10.2	0.3
5a	Caohai water level	1.5 m	25.3	15.0	23.3	16.8	0.2
5b		3.5 m	25.3	14.1	21.4	17.2	0.8

3.3.4 Caohai Shallow Lake Model Results

The simulation of the water temperature by the standard StoLaM run was compared to the temperature measurement from the Kunming Institute of Environmental Sciences (see Fig. 3.6). At the Caohai, there were two temperature-monitoring stations. One station was situated in the middle of the lake, the other one called Duanqiao. The good

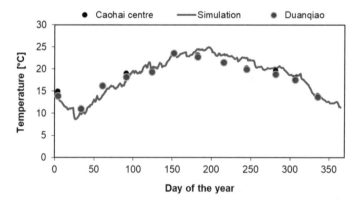

Fig. 3.6 Simulation of the water temperature in 2016 (lines: simulation result. Black dots: Caohai Centre monitoring station; red dots: Duanqiao monitoring station)

agreement between simulation and measurements is an indication that the hydrological and hydrodynamic conditions for the Lake Caohai can be well reproduced with the model.

Standard Scenario

The comparison of the results from the simulation with the measured values from the Kunming Environmental Science Institute showed that the standard simulation run fits the field data well while taking the measured nutrient sediment release data into consideration (scenario 1, compare Table 3.9). In this scenario, the dynamics of the cyanobacteria *Microcystis* can also be described without further food web interactions.

Sediment Nutrient Release Scenario

The hypothetical scenario without sediment release shows a massive reduction of the algal bloom and the phosphorus concentration in summer compared to the standard run (scenario 2a). This illustrates the high relevance of the internal lake phosphorus release during summer in this eutrophic shallow lake. As expected, a reduction of the sediment area by 25% causes only a smaller improvement of the water quality for all parameters (scenario 2b).

Tributary Water Quality Scenario

Reducing the nutrient content of the inflow by 100% (scenario 3a) clearly reduces the total concentration of algae and phosphorus in the lake, but will not change their overall dynamics in the short term due to the very high internal P-loading by the sediments. Only a clear decrease in nitrogen concentrations is observable. The same is true for the scenario 3b with only 50% of the nutrient loading, but the effect is much smaller than in scenario 3a.

Tributary Water Quantity Scenario

The use of annual mean values for the discharge of the tributaries (scenario 4a) reduces the algal and phosphorus peaks in summer, as the measured reduced inflow in summer was artificially compensated in the model, but does not have a fundamentally positive effect on nutrient concentrations. Nevertheless, this shows the clear influence of the inflow fluctuations of the Daguan River on the water quality of the entire lake. Doubling the water inflow rate significantly reduces the algae and phosphorus content during summer in the lake due to leaching despite the unchanged high nutrient concentrations in the inflow (scenario 4b). Halving, on the other hand, does not make much of a difference, as the positive effect of less nutrients is compensated by less washout of algae and phosphorus (Scenario 4c).

Caohai Water Level Scenario

Lowering the water depth by 1 m increases the algal bloom considerably due to the lower water volume (scenario 5a), especially since the nutrients released from the sediment concentrate more quickly in the water column. In comparison, increasing the water depth by 1 m causes only a relatively small change in nutrient concentration,

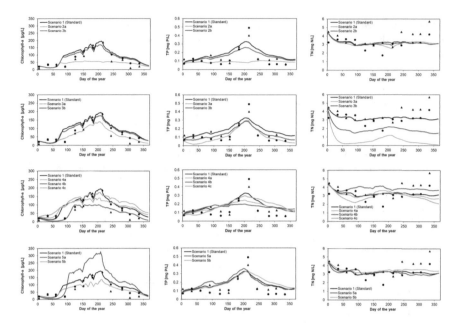

Fig. 3.7 Results of the scenario simulation with StoLaM (lines: simulation results. dots: Caohai Centre monitoring station; triangles: Duanqiao monitoring station)

but reduced the cyanobacterial biomass significantly since a deeper vertical mixing into the water column causes a poorer light supply to the algae (scenario 5b) (Fig. 3.7).

In addition to the nutrient concentrations over the year, mass balances also provide important information on enrichment or depletion in the water body.

In the case of the phosphorus budget, it can be seen that the scenarios only slightly differ in the whole-year balance and usually reach the low spring phosphorus level again in the subsequent winter, even if there is sometimes a slight increase in the water column (see balance column in Table 3.5). For the scenario with doubled inflow (4b), the higher nutrient inflow from the tributaries results in slightly higher water column enrichment than in the standard run (scenario 1). Furthermore, the enrichment is higher in the scenario with a water depth of 3.5 m (5b) due to a overall slightly reduced phosphorus sedimentation. Only in the scenario without nutrient input from the rivers (3a) phosphorus depletion in the water column can be found compared to the starting conditions of the simulated year.

Looking at the difference between nutrient release from the sediment and sedimentation rates (Table 3.5), the hypothetical scenario 3a shows a depletion of the phosphorus content in the sediment, as less particulate phosphorus enter the lake by the inflows. In most other scenarios, however, sedimentation exceeds the sediment release and leads to a long-term accumulation of phosphorus in the sediment.

For nitrogen, the picture is slightly different (Table 3.10). Here the nitrogen whole-year balance for the water column of the status quo scenario 1 shows already a

Table 3.10 Simulated nitrogen flows and balance (yearly average) in mg N m^{-2} d^{-1}

No	Description of scenario	Sub-scenario	Inflow	Sediment release	Sedimentation	Outflow	Denitrification	Balance
1	Standard		638.7	63.8	385.7	321.6	4.1	−8.9
2a	Sediment nutrient release	No release	638.7	–	337.7	301.7	8.3	−9
2b		75% release (= 25% lake area dredged)	638.7	48.4	378.0	313.7	4.2	−8.8
3a	Tributary water quality	0% nutrient inflow	–	67.7	37.3	57.8	0.6	−28
3b		50% nutrient inflow	320.6	64.7	217.3	185.9	1.8	−19.7
4a	Tributary water quantity	Continuous average	645.0	62.5	391.2	321.9	2.9	−8.5
4b		200% inflow	1290.0	53.9	599.7	735.9	9.6	−1.3
4c		50% inflow	338.7	67.4	261.1	157.7	1.9	−14.6
5a	Caohai water level	1.5 m	638.7	65.6	398.2	306.4	5.1	−5.4
5b		3.5 m	638.7	61.5	359.8	344.7	4.9	−9.2

washout of 8.9 mg m^{-2}d^{-1}, which is in the same order of magnitude as the other scenarios.

Denitrification of dissolved inorganic nitrogen plays only a minor role in the nitrogen balance of the lake.

The low level of denitrification in the water body is a result of the assumption, based on the available measurements, that a large part of the inflowing nitrogen is present in particulate form as detritus bound and sinks relatively quickly onto the sediment.

Overall, the available simulations permit the conclusion that nitrogen also accumulates massively in the sediment, which should be verified by further measurements.

3.4 Summary and Recommendations for the Dianchi Shallow Lake Management

3.4.1 Shallow Lake Management – A Very Special Task

Shallow lake management is even more challenging than the management of deep-water lakes. Shallow lakes are more vulnerable than deeper waters due to a higher ratio between sediment area and water volume as well as lower retention capacity for nutrients, and because interactions between pollution, physical and ecological components are linked with a high dynamic. Because of the sensitivity of shallow lakes, shallow lake management must follow a very consistent and strict management strategy.

With the 2018 Revision of World Urbanization Prospects the UN (UN, 2019) points out "Sustainable urbanization is key to successful development. Understanding the key trends in urbanization likely to unfold over the coming years is crucial to the implementation of the 2030 Agenda for Sustainable Development, including efforts to forge a new framework of urban development." The poor status of Dianchi points to an overwhelmingly obvious example of cause and effect in that context. Unsustainable growth of Kunming caused a level of pollution for the lake, which exceeded the carrying capacity of that very vulnerable water resource.

The remediation of Dianchi will take a full and lasting commitment in a water and resources management based on a system understanding of cause and effect for the benefit of Kunming. Thus, "the future growth of cities and concomitant appropriation of land and natural resources will determine success towards an environmentally sustainable future. In some cities, unplanned or inadequately managed urban expansion leads to rapid sprawl, pollution, and environmental degradation." (UN, 2019).

Today's pressures on Dianchi are much more complex and severe compared to those of ancient China. Modern pressures need integral solutions. Thus, the Dianchi Management Authorities have agreed on six main measures consisting of (1) waste water treatment & circular sewer, (2) lakeside remediation, (3) river remediation, (4) sediment management, (5) conservation of water resources, and (6) water transfer/efficiency. During our SINOWATER studies and together with our partners from Kunming and experts from other parts of China we discussed the relevance of evaluation as an instrument for water management. An economic benefit approach for example indicated that only subtle further investment and easy technical modifications could produce relevant additional purification effects for the lakeside wetlands (Wermter [41]).

3.4.2 Necessity for Highest Data Quality

The intention of the survey of the water quality of the northern part of the lake called Caohai was to produce the basis for the system understanding of the shallow lake of Caohai. Above we described our two step approach. First we collected available environmental, hydrological and pollution data from Kunming authorities and other sources. It was possible to parameterise a lake model based on the preliminary data set. It supported first estimations of management scenarios. But the first and quick answers came with a relative vagueness. Relevant monitoring data was still missing like P release from lake-internal sediment. One of the less astounding but nevertheless crucial findings of this very practical approaches was the acknowledgement of the necessity for a sufficiently high data quality, both quantitatively and qualitatively.

Because data uncertainty causes model uncertainty and model uncertainty causes planning and management uncertainties, we recommended and conducted a thorough sediment and river pollution monitoring campaign in 2017. Examination of the sediment phosphorus release and additional measurements of N and P was crucial for the improvement of the shallow lake model of Caohai. In May 2018 we presented and discussed the results of StoLaM shallow lake model of Caohai and the out-comes of a set of management scenarios.

In general, the Stoichiometric Lake Model StoLaM is applicable for the modelling of Lake Caohai and provides evaluable results for the selected scenarios. The model results clearly show that with the measured nutrient sediment release rates, the seasonal dynamics of the water quality parameters in the water column, measured by the Kunming Institute for Environmental Sciences, can be well simulated. Nevertheless, some measurements are still subject to increased uncertainty. The knowledge of the particle-bound nutrient content in the tributaries and its sedimentation and remineralisation rates are crucial for a quantitatively correct analysis and modelling of the water quality in Lake Caohai, and should be further investigated. Also, the very high nutrient release rates of the sediments under both, oxic and anoxic conditions, should be confirmed by further measurements. Besides, more measurements of the water inflows are necessary in future to have a broader data basis, especially in the River Daguan.

There are also factors, which were still unknown. Meanwhile, we have an estimation of the nutrient release from the sediment, but there is a lack on data or estimations of the nutrient release by resuspended detritus particles due to lake internal processes such as wind-induced turbulence. These resuspension events might have also a significant potential for nutrient release into the water column.

3.4.3 Recommendations of Most Promising Management Scenario for the Lake Management

Based on the simulation results in this project we can derive clear recommendations for the lake management of Lake Caohai. We already saw that decentralised sponge city measures in the catchment area could improve the nutrient concentration in the lake and thus the potential for washout. Especially scenarios with an increased water quantity and a reduced nutrient loading in the lake inflow would have a huge impact. However, dropping the discharge to an amount as shown in the simulated scenarios would not be possible by sponge city measures in the next years only.

Summing up, we recommend to improve the trophic situation of the rivers flowing into Lake Caohai to prevent huge lake nutrient intake and to improve the washing out capacity. A significant reduction in the nutrient load of the tributaries can improve the lake water quality immensely. However, dissolved and particulate nutrient fractions should be evaluated separately. In general, an increase of inflowing water quantities favours an improvement of the water quality in the lake through wash-out processes, especially during the summer months.

Changes in the filling level of this shallow lake can have significant positive and negative effects on the water quality by changing the water retention time and the sediment area/water volume ratio. In any case, a significant increase in the mean water depth of the lake would have a clearly positive effect on the reduction of algal blooms.

We also recommend a significant reduction of the nutrient release rate in the lake, which would also greatly reduce the nutrient concentration in the summer months. According to our data, this can only be achieved by dredging, as there are massive nutrient releases even under aerobic conditions.

The measures discussed here (deepening of the lake, increased inflow, reduced nutrient load, reduced sediment release) can be combined without any conflict of optimisation for the lake water quality. The efficiency of different combinations can also be assessed with the modelling approach presented here.

We also recommend to apply the StoLaM model or a comparable model to the Dongfengba part of Lake Caohai, as the monitoring results show that the water quality in this part and its tributaries is much worse than in the eastern part of the Caohai.

Annex

Annex 3.1: Full Compares Between Aerobic and Anoxic Milieu

Aerobic

		24.07.2017			25.07.2017			26.07.2017			27.07.2017			28.07.2017			29.07.2017			30.07.2017			31.07.2017			01.08.2017			02.08.2017			03.08.2017			04.08.2017		
		O₂	Rdx	Temp	O₂	Rdx	Temp	O₂	Rdx	Temp	O₂	Rdx	Temp	O₂	Rdx	Temp	O₂	Rdx	Temp	O₂	Rdx	Temp	O₂	Rdx	Temp	O₂	Rdx	Temp	O₂	Rdx	Temp	O₂	Rdx	Temp	O₂	Rdx	Temp
1	May	7,1	53	22,0	7,0	119	21,6	7,1	90	21,6	7,1	88	22,1	7,0	101	21,4	7,0	99	22,2	7,1	102	21,8	7,0	109	22,1	7,0	105	22,3	7,0	107	21,5	7,0	103	21,7	7,0	102	22,2
	July	7,3	49	22,0	7,3	113	21,6	6,9	105	21,6	7,0	114	22,1	7,0	111	21,4	7,1	103	22,2	7,0	107	21,8	7,1	104	22,1	7,1	109	22,3	7,1	106	21,5	7,1	99	21,7	7,2	110	22,2
6	May	7,1	52	22,0	7,1	116	21,6	7,0	93	21,6	7,0	92	22,1	7,0	100	21,4	7,0	109	22,2	7,0	103	21,8	7,1	106	22,1	7,1	100	22,3	7,0	101	21,5	7,0	112	21,7	7,0	111	22,2
	July	7,2	52	22,0	7,3	115	21,6	7,1	113	21,6	7,1	105	22,1	7,0	108	21,4	7,0	108	22,2	7,0	112	21,8	7,1	111	22,1	7,1	102	22,3	7,0	105	21,5	7,0	115	21,7	7,0	108	22,2
7	May	7,1	48	22,0	7,1	111	21,6	7,1	112	21,6	7,1	96	21,8	7,0	102	21,4	7,0	96	22,2	7,1	100	21,8	7,0	102	22,1	7,0	106	22,3	7,1	110	21,5	7,1	110	21,7	7,0	104	22,2
	July	7,2	50	22,0	7,2	114	21,6	7,0	114	21,6	7,1	96	22,1	7,1	92	21,4	7,0	96	22,2	7,0	99	21,8	7,1	104	22,1	7,1	104	22,3	7,1	100	21,5	7,0	105	21,7	7,0	103	22,2
8	May	7,2	45	22,0	7,3	110	21,6	7,3	89	21,6	7,1	96	21,8	7,0	96	21,4	7,1	111	22,2	7,1	112	21,8	7,1	111	22,1	7,1	115	22,3	7,1	110	21,5	7,1	111	21,7	7,1	108	22,2
	July	7,2	51	22,0	7,1	115	21,6	7,0	103	21,6	7,0	101	22,1	7,1	96	21,4	7,1	102	22,2	7,1	105	21,8	7,1	98	22,1	7,1	96	22,3	7,0	102	21,5	7,0	111	21,7	7,1	111	22,2
	Blind	7,1	114	22,0	7,2	49	21,6	7,2	86	21,6	7,1	92	22,1	7,1	89	21,4	7,1	90	22,2	7,1	100	21,8	7,1	105	22,1	7,1	104	22,3	7,1	114	21,5	7,1	113	21,7	7,1	112	22,2

Anaerobic

		24.07.2017			25.07.2017			26.07.2017			27.07.2017			28.07.2017			29.07.2017			30.07.2017			31.07.2017			01.08.2017			02.08.2017			03.08.2017			04.08.2017		
		O₂	Rdx	Temp	O₂	Rdx	Temp	O₂	Rdx	Temp	O₂	Rdx	Temp	O₂	Rdx	Temp	O₂	Rdx	Temp	O₂	Rdx	Temp	O₂	Rdx	Temp	O₂	Rdx	Temp	O₂	Rdx	Temp	O₂	Rdx	Temp	O₂	Rdx	Temp
1	May	0,7	-105	22,0	0,2	-118	21,6	0,2	-104	21,6	0,2	-111	22,1	0,3	-105	21,4	0,3	-99	22,2	0,3	-105	21,8	0,2	-98	22,1	0,2	-101	22,3	0,2	-96	21,5	0,2	-103	21,7	0,1	-105	22,2
	July	0,6	-100	22,0	0,2	-63	21,6	0,3	-76	21,6	0,2	-80	22,1	0,2	-85	21,4	0,2	-99	22,2	0,2	-110	21,8	0,2	-105	22,1	0,2	-104	22,3	0,2	-99	21,5	0,2	-100	21,7	0,1	-107	22,2
6	May	0,6	-150	22,0	0,6	-145	21,6	0,7	-123	21,6	0,6	-115	22,1	0,6	-99	21,4	0,4	-86	22,2	0,4	-100	21,8	0,3	-111	22,1	0,3	-109	22,3	0,2	-107	21,5	0,2	-105	21,7	0,2	-99	22,2
	July	0,5	-14	22,0	0,4	-64	21,6	0,3	-81	21,6	0,3	-99	22,1	0,3	-95	21,4	0,3	-98	22,2	0,2	-93	21,8	0,2	-98	22,1	0,2	-99	22,3	0,2	-108	21,5	0,1	-103	21,7	0,1	-98	22,2
7	May	0,5	-130	22,0	0,2	-134	21,6	0,3	-89	21,6	0,3	-93	22,1	0,2	-103	21,4	0,1	-110	22,2	0,2	-105	21,8	0,2	-103	22,1	0,1	-103	22,3	0,1	-105	21,5	0,1	-102	21,7	0,1	-100	22,2
	July	0,4	-95	22,0	0,6	-86	21,6	0,6	-85	21,6	0,5	-83	22,1	0,5	-89	21,4	0,4	-90	22,2	0,3	-99	21,8	0,3	-108	22,1	0,2	-102	22,3	0,2	-110	21,5	0,2	-119	21,7	0,2	-112	22,2
8	May	0,6	-121	22,0	0,2	-98	21,6	0,3	-94	21,6	0,2	-27	22,1	0,2	-90	21,4	0,2	-99	22,2	0,2	-96	21,8	0,2	-89	22,1	0,2	-99	22,3	0,2	-105	21,5	0,2	-431	21,7	0,1	-110	22,2
	July	0,5	-96	22,0	0,3	-106	21,6	0,2	-102	21,6	0,2	-112	22,1	0,3	-101	21,4	0,2	-115	22,2	0,2	-103	21,8	0,2	-97	22,1	0,3	-105	22,3	0,3	-113	21,5	0,2	-105	21,7	0,5	-103	22,2
	Blind	4,8	118	22,0	5,6	93	21,6	5,8	87	21,6	5,7	86	21,8	5,7	82	21,4	6,0	90	22,2	5,7	89	21,8	5,7	91	22,1	5,6	87	22,3	5,6	93	21,5	5,5	93	21,7	5,3	93	22,2

Oxgen in mg/l, redoxpotential in mV, Temperature in °C, hatchure for values outside of tolerance

Annex 3.2: Relevant Parameters and Variables of StoLaM Used in the Lake Caohai Simulation Study

For further details on model structure, equations and parameters see Strauss [29].

Parameter	Value	Unit
Sedimentation rate of detritus	1	$m \cdot day^{-1}$
Temperature coefficient of sediment nutrient release	2.4227	–
Sediment phosphate release rate (20 °C)	19*	$mg\ PO_4\text{-}P \cdot m^{-2} \cdot day^{-1}$
Sediment ammonium release rate (20 °C)	83*	$mg\ NH_4\text{-}P \cdot m^{-2} \cdot day^{-1}$
Denitrification coefficient (20 °C)	10**	$L \cdot m^{-2} \cdot day^{-1}$
Temperature coefficient of denitrification	1.02	–
Nitrification rate from ammonium to nitrate (20 °C)	0.124	day^{-1}
Temperature coefficient of nitrification	1.065	–
Degradation rate of detritus (at 20 °C)	0.2**	day^{-1}
Remineralisation rate of detrital phosphorus (at 20 °C)	0.2**	day^{-1}
Remineralisation rate of detrital nitrogen (at 20 °C)	0.25**	day^{-1}
Remineralisation rate of DOP (at 20 °C)	0.02	day^{-1}
Remineralisation rate of DON (at 20 °C)	0.02	day^{-1}
Temperature coefficient of remineralization	1.13	–
Light attenuation coefficient for water and DOM	0.3	m^{-1}
Light attenuation coefficient for detritus	0.016	$L \cdot m^{-1} \cdot mg\ ww^{-1}$
Light attenuation coefficient for blue-green algae	0.012	$L \cdot m^{-1} \cdot \mu g\ Chl\text{-}a^{-1}$
Chlorophyll-a content of blue-green algae	5.13	$\mu g\ Chla \cdot mg\ ww\ Phyto^{-1}$
Maximum potential growth rate of *Microcystis*	1.5	day^{-1}
Maximum natural mortality of *Microcystis*	0.1	day^{-1}
Respiration rate (at T_{opt}) of *Microcystis*	0.16	day^{-1}
Sedimentation rate of *Microcystis* (buoyancy)	−1.38	$m \cdot day^{-1}$
Temperature coefficient of respiration	1.13	–
Constant of light limitation (PhAR; Jassby & Platt, !976) of *Microcystis*	25	$J \cdot m^{-2} \cdot s^{-1}$
Temperature maximum *Microcystis*	35	°C
Temperature minimum *Microcystis*	0	°C
Temperature optimum *Microcystis*	30	°C
Maximum phosphorus cell quota of *Microcystis*	0.004	$mg\ P \cdot mg\ ww$
Minimum phosphorus cell quota of *Microcystis*	0.0015	$mg\ P \cdot mg\ ww$
Maximum nitrogen cell quota of *Microcystis*	0.06	$mg\ N \cdot mg\ ww$
Minimum nitrogen cell quota of *Microcystis*	0.011	$mg\ N \cdot mg\ ww$
Half saturation constant for phosphorus uptake of *Microcystis*	0.0006	$mg\ P \cdot L^{-1}$

(continued)

(continued)

Parameter	Value	Unit
Half saturation constant for nitrogen uptake of *Microcystis*	0.03	mg N $\cdot$ L^{-1}
Maximum phosphorus uptake rate of *Microcystis*	0.12	mg P $\cdot$ mg ww^{-1} $\cdot$ day^{-1}
Maximum nitrogen uptake rate of *Microcystis*	0.1	mg N $\cdot$ mg ww^{-1} $\cdot$ day^{-1}

*: Parameter measured in this study. **: Parameter calibrated in this study.

References

1. Bloesch J, Schroeder HG (2008) Integrated transboundary management of lake constance driven by the international commission for the protection of lake constance (IGKB). In: Nato science series. Integrated water management - practical experiences and case studies. Springer, Netherlands
2. Cao X, Song C, Xiao J, Zhou Y (2018) The optimal width and mechanism of riparian buffers for storm water nutrient removal in the Chinese eutrophic lake Chaohu watershed. Water 10(10):1489
3. Chen K, Disse M (2021) Development of innovative 3D water governance models for Kunming City and Bavaria. In: Chinese water systems, vol 4. Springer, Heidelberg
4. de Groot R, Brander L, van der Ploeg S, Costanza R, Bernard F, Braat L, Christie M, Crossman N, Ghermandi A, Hein L, Hussain S (2012) Global estimates of the value of ecosystems and their services in monetary units. Ecosyst Serv 1(1):50–61. https://doi.org/10.1016/j.ecoser.2012.07.005
5. Ding H, Zhang Q, Sun R, Su L 丁煌英 张庆 孙鲁夫 苏兰英 (2012) Assessment of water quality in dry season of Daguan River in Kunming by using plankton community structure characteristics 用浮游生物群落结构特征评价昆明大观河枯水期水质. J Southwest For Univ 32(6):83–87. https://doi.org/10.3969/jissn2095-1914201206017
6. Diovisalvi N, Bohn VY, Piccolo MC, Perillo GME, Baigún C, Zagarese HE (2014) Shallow lakes from the Central Plains of Argentina: an overview and worldwide comparative analysis of their basic limnological features. Hydrobiologia 752(1):5–20. https://doi.org/10.1007/s10750-014-1946-x
7. Grambow M (2008) Wasser management Integriertes Wasser-Ressource nmanagement von der Theorie zur Umsetzung. Vieweg Heidelberg
8. Gross G (2017) Some effects of water bodies on the n environment–numerical experiments. J Heat Island Inst Intern 12(2):1–11
9. Han M, Gong Y, Zhou C, Zhang J, Wang Z, Ning K (2016) Comparison and interpretation of taxonomical structure of bacterial communities in two types of lakes on Yun-Gui plateau of China. Sci Rep 6:30616.https://doi.org/10.1038/srep30616
10. Hargeby A, Blindow I, Hansson L-A (2004) Shifts between clear and turbid states in a shallow lake: multi-causal stress from climate, nutrients and biotic interactions. Arch Hydrobiol 161(4):433–454. https://doi.org/10.1127/0003-9136/2004/0161-0433
11. He J (2018) Water quality target management and pollution load control in Dianchi Lake basin. Presented at the 2nd Sino-German symposium on integrated lake basin management, Dianchi, Kunming, 7–8 May 2018
12. He J, Wu X, Zhang Y, Zheng B, Meng D, Zhou H, Lu L, Deng W, Shao Z, Qin Y (2020) Management of water quality targets based on river-lake water quality response relationships for lake basins – a case study of Dianchi Lake. Environ Res 186(2020):1–16. https://doi.org/10.1016/j.envres.2020.109479

13. Huang J, Zhang Y, Huang Qi, Gao J (2018) When and where to reduce nutrient for controlling harmful algal blooms in large eutrophic lake Chaohu, China? Ecol Indic 89:808–817. https://doi.org/10.1016/j.ecolind.2018.01.056

14. Huang K, Jin Z, Li J, Yang F, Zhou B (2014) Technologies and engineering practices of water pollution control for a typical urban river in Dianchi watershed. Fresenius Environ Bull 23:3469–3475

15. Huang K 黄可, Zhang X 张先智, Zhang H 张恒明, Jin Z 金竹静, Li J 李金花, Kong D 孔德平, Zhou B 周保学, Yang F 杨逢乐, Chai J 柴金岭 (2015) Effect of a restoration project on zooplankton and benthic animals in Xinyunliang River 治理工程对新运粮河浮游动物与底栖动物影响. Environ Sci Technol 38(10):11–15

16. Jin XC, Wang L, He LP (2006) Lake Dianchi: experience and lessons learned brief. In: International lake environment committee (edn): lake basin management initiative: final main report, Paris

17. Jørgensen SE (2010) A review of recent developments in lake modelling. Ecol Model 221:689–692. https://doi.org/10.1016/j.ecolmodel.2009.10.022

18. Kunming Statistics Bureau 昆明市统计局 (2018) Statistical communique of national economic and social development of Kunming in 2017 2017 年昆明市国民经济和社会发展统计公报. http://wwwkmgovcn/c/2018-05-10/2571823s.html. Accessed 7 May 2019

19. Lei X, Hua L, Xinqiang L, Yuxin Y, Li Z, Xinyi C (2012) Water quality parameters response to temperature change in small shallow lakes. Phys Chem Earth Parts A/B/C 47–48:128–134. https://doi.org/10.1016/jpce201011005

20. Li Z, Xie P, Du P (2016) Simulation of pollution control in Laoyunliang river Kunming. Acta Sci Circumst 37:1657–1667

21. Küppers S, Yin D, Zheng B, Tiehm A (2019) Fostering water treatment in eutrophic areas: innovative water quality monitoring and technologies mitigating taste & odor problems demonstrated at Tai Hu. In: Köster S, Reese M, Zuo J (eds.) Urban water management for future cities future city, vol 12, Springer, Cham

22. Ma X 马勋静, Hu K 胡开林, Yan X 严雪梅, Yu M 余美, Feng C 冯成义 (2010) Summary of wastewater treatment in Wujiadui constructed wetland 五家堆人工湿地处理污水的工程总结. China Water Wastewater 26(14):96–98

23. Moore PA, Reddy KR, Fisher MM (1998) Phosphorus flux between sediment and overlying water in lake Okeechobee, Florida: spatial and temporal variations. J Environ Qual 27(6):1428–1439. https://doi.org/10.2134/jeq1998.00472425002700060020x

24. Nie J 聂菊芬 (2015) Water quality monitoring in the lower reaches of Xinyunliang river 新运粮河下游水质监测. Environ Sci Surv 1:12–14. https://doi.org/10.3969/jissn1673-965520 1501004

25. Nixdorf E, Zhou C (2019) Background information about Poyang Lake basin. In: Yue T et al (eds) Chinese water systems terrestrial environmental sciences. Springer, Cham

26. Qin B, Yang G, Ma J, Wu T, Li W, Liu L, Deng J, Zhou J (2018) Spatiotemporal changes of cyanobacterial bloom in large shallow eutrophic lake Taihu, China. Fronti Microbiol 9:451. https://doi.org/10.3389/fmicb201800451

27. Qinghui Z, Xuhui D, Yuwei C, Xiangdong Y, Min X Davidson T, Jeppesen E (2018) Hydrological alterations as the major driver on environmental change in a floodplain Lake Poyang (China): evidence from monitoring and sediment records. J Great Lakes Res 44(3):377–387. https://doi.org/10.1016/jjglr201802003

28. Ryan PJ, Stolzenbach KD (1972) Engineering aspects of heat disposal from power generation. In: Harleman DRF (ed) M Parson laboratory for water resources and hydrodynamics. Department of civil engineering, Massachusetts Institute of Technology, Cambridge, MA

29. Strauss T (2009) Dynamische Simulation der Planktonentwicklung und interner Stoffflüsse in einem eutrophen Flachsee [Simulation of plankton dynamics and internal nutrient fluxes in a eutrophic shallow lake]. Ph.D. Thesis RWTH Aachen University Shaker, Aachen. https://doi.org/10.18154/RWTH-CONV-113654

30. Strauss T, Gabsi F, Hammers-Wirtz M, Thorbek P, Preuss TG (2017) The power of hybrid modelling: an example from aquatic ecosystems. Ecol Model 364:77–88. https://doi.org/10.1016/j.ecolmodel.2017.09.019

31. Su L, Lu J, Christensen P (2010) Comparative study of water resource management policies between China and Denmark. Procedia Environ Sci 2:1775–1798
32. United Nations, Department of Economic and Social Affairs, Population Division (2019) World Urbanization Prospects: The 2018 Revision (ST/ESA/SERA/420). United Nations, New York
33. Vadeboncoeur Y, Jeppsen E, Zanden MJV, Schierup H, Christoffersen K, Lodge DM (2003) From Greenald to green lakes: cultural eutrophication and the loss of benthic pathways in lakes. Limnol Oceanogr 48:1408–1418. https://doi.org/10.4319/lo20034841408
34. Vinçon-Leite B, Casenave C (2019) Modelling eutrophication in lake ecosystems: a review. Sci Total Environ 651:2985–3001. https://doi.org/10.1016/j.scitotenv.2018.09.320
35. Wang Y, Wermter R (2021) The planning, management and decision support systems of Kunming's urban drainage system. In: Chinese water systems, vol 4. Springer, Heidelberg
36. Wang H 王华光, Liu B 刘碧波, Li X 李小平, Liu J 刘剑彤, Ao H 敖鸿毅, Li Q 李清曼(2012) Seasonal variation of water quality of Xinyunliang river in Dianchi lake and the effect of ecological restoration in the riparian zone 滇池新运粮河水质季节变化及河岸带生态修复的影响. J Lake Sci 24(3):334–340. https://doi.org/10.18307/20120302
37. Wang Y, Weiping Hu, Peng Z, Zeng Ye, Rinke K (2018) Predicting lake eutrophication responses to multiple scenarios of lake restoration: a three-dimensional modeling approach. Water 10(8):994. https://doi.org/10.3390/w10080994
38. Wang Z, Kai H, Peijiang Z, Huaicheng G (2010) A hybrid neural network model for cyanobacteria bloom in Dianchi Lake. Procedia Environ Sci 2:67–75. https://doi.org/10.1016/jproenv20 1010010
39. Wang Z, Zhang Z, Zhang Y, Zhang J, Yan S, Guo J (2013) Nitrogen removal from Lake Caohai, a typical ultra-eutrophic lake in China with large scale confined growth of Eichhornia crassipes. Chemosphere 92(2):177–183. https://doi.org/10.1016/j.chemosphere.2013.03.014
40. Yali Wu, Bo Y, Zhou F, Tang Q, Guimberteau M, Ciais P, Yang T, Peng S, Piao S, Zheng J, Dong Y, Dai C (2018) Quantifying the unauthorized lake water withdrawals and their impacts on the water budget of eutrophic lake Dianchi, China. J Hydrol 565:39–48. https://doi.org/10. 1016/j.jhydrol.2018.08.017
41. Wermter P (2018) Evaluation of measures using the example of constructed wetlands at the Dianchi Shore. Presented at the 2nd sino-german symposium on integrated lake basin management Dianchi, Kunming, 7–8 May 2018
42. Xie F, Li L, Song K, Li G, Wu F, Giesy JP (2019) Characterization of phosphorus forms in a Eutrophic Lake China. Sci Total Environ 659:437–1447.https://doi.org/10.1016/j.scitotenv. 2018.12.466
43. Yan K, Yuan Z, Goldberg S, Gaoe W, Ostermann A, Xu J, Zhang F, Elser J (2019) Phosphorus mitigation remains critical in water protection: a review and meta-analysis from one of China's most eutrophicated lakes. Sci Total Environ 689:1336–1347. https://doi.org/10.1016/j.scitot env.2019.06.302
44. Yang F 杨逢乐, Jin Z 金竹静 (2008) Present situation of water environment pollution of rivers in the North Bank of Dianchi lake and countermeasures 滇池北岸河流水环境污染现状及防治对策研究. Environ Sci Guide 环境科学导刊 27(6):43–46. https://doi.org/10.13623/jcnkihkdk200806012
45. Yang J 杨锦凤 (2018) Analysis of regulation effects at Wulong river 乌龙河整治效果分析. Environ Sci Surv 37(3):25–29
46. Li Y, Wei J, Gao X, Chen D, Weng S, Wei Du, Wang W, Wang J, Tang C, Zhang S (2018) Turbulent bursting and sediment resuspension in hyper-eutrophic Lake Taihu, China. J Hydrol 565:581–588. https://doi.org/10.1016/j.jhydrol.2018.08.067
47. Yua Y, Guana J, Maa Y, Yub S, Guoc H, Baoa L (2011) Aquatic environmental quality variation in Lake Dianchi watershed. Procedia Environ Sci 2:76–81
48. Zhan J 展巨宏, Li H 李辉, Li Q 李强, Zhao S 赵世民, Yang Y杨彦臻, Pan X 潘学军(2014) Effect of comprehensive river sewage closure treatment project on water quality in Chuanfang river Kunming 河流综合截污工程对昆明船房河水质的改善效果. Environ Chem 33(1):148–153. https://doi.org/10.7524/jissn0254-6108201401002

49. Zhiyi L, Pencheng X, Pengfei D (2017) Simulation of pollution control in Laoyunliang river Kunming. Acta Sci Circum 37(5):1657–1667. https://doi.org/10.13671/jhjkxxb20160403
50. Zhang L, Liu J, Zhang D, Luo L, Liao Q, Yuan L, Wu N (2018) Seasonal and spatial variations of microcystins in Poyang Lake the largest freshwater lake in China. Environ Sci Pollut Res 25:6300–6307. https://doi.org/10.1007/s11356-017-0967-1
51. Zhang S, Wang W, Zhang K, Peiyao Xu, Yin Lu (2018) Phosphorus release from cyanobacterial blooms during their decline period in eutrophic Dianchi Lake, China. Environ Sci Pollut Res 25(14):13579–13588. https://doi.org/10.1007/s11356-018-1517-1
52. Zhu R, Wang H, Chen J, Shen H, Deng X (2017) Use the predictive models to explore the key factors affecting phytoplankton succession in Lake Erhai, China. Environ Sci Pollut Res 25(2):1283–1293. https://doi.org/10.1007/s11356-017-0512-2

Chapter 4
New Technical Approaches for the Co-processing of Pharmaceutical Wastewater in Municipal Wastewater Treatment Plants in the Shenyang Region

Yunbo Yun and Max Dohmann

4.1 Introduction

Hardly any other country has experienced such immense and rapid growth in recent years as the People's Republic of China. In the years 2000–2010, the annual growth rate of the gross domestic product was about 10%; in the last few years (2012–2017), it slowly slowed down to about 7% (World Bank 2015). A major contribution to these rapid developments has been made by the constant growth of industry, which at the same time is leading to much greater environmental problems, including water pollution. The pollution of water resources results mainly from municipal and industrial wastewater discharges. Thus, in addition to unsatisfactory conditions, especially of surface waters, there are often problems in the safe use of these resources for the water supply of the population, industry and agriculture.

The research project SINOWATER- "Good Water Governance Management and innovative technologies to improve water quality in two important Chinese waters" was started in the course of the environmental policy of the Chinese government. With the 11th Chinese five-year plan, which came into force in 2006, concrete targets for improving water quality were set for the first time. In order to achieve the goals set, a water programme, the "National Major Program of Science and Technology for Water Pollution Control and Governance", was established under the leadership of the Chinese Ministry of Environment and the involvement of six other Chinese ministries of the central government. This programme aims to improve water quality and ecological conditions in large Chinese river basins and lakes, which are to be achieved through mega water projects. The time frame for achieving the objectives has been set for the period between 2006 and 2020. The Federal Ministry of Education and Research (BMBF) concluded an agreement with the Chinese Ministry of Science

Y. Yun · M. Dohmann (✉)
Research Institute for Water and Waste Management at the RWTH Aachen University (FiW) e. V.,
Aachen, Germany

© The Author(s) 2022 103
M. Dohmann et al. (eds.), *Chinese Water Systems*, Terrestrial Environmental Sciences,
https://doi.org/10.1007/978-3-030-80234-9_4

and Technology (MoST) in 2012, which provides for the participation of German partners in the research projects of the mega water projects.

In 2015, the BMBF approved the funding application for the research project SINOWATER- "Good Water Governance Management and innovative technologies for improving water quality in two significant Chinese waters". The results of the coordination with the responsible persons of two mega water projects led to concentration on one river and one lake catchment area each, namely the Liao River in the province of Liaoning in the northeast of China and the Dianchi Lake near Kunming in the southwest of China.

In the project description for the SINOWATER project the following four main objectives are formulated FIW E.V. [5]:

1. "Development of changed structures and organisational measures for improved analysis and decision-making in the normative and operational management of the water sector on the basis of cooperative, participatory and specific ecological research approaches".
2. "Reduction of environmental pollution of water bodies from municipal and industrial wastewater treatment plants".
3. "Participation in the updating of a long-term master plan for Dianchi Lake."
4. "Development of a sustainable concept for sewage sludge disposal in the Shenyang region.

Based on the main objectives, a total of 7 sub-projects are planned under the overall project. Figure 4.1 shows the allocation of the seven sub-projects to the two mega water projects. The whole project is divided into management (M), concept (K) and

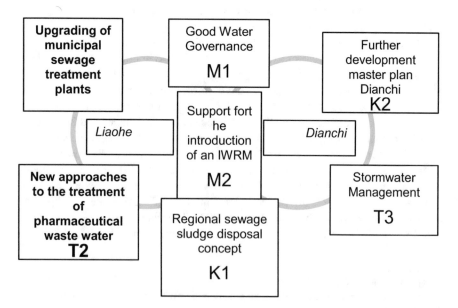

Fig. 4.1 Relationship between the Chinese mega water projects and SINOWATER sub-projects

technology projects (T). The two sub-projects T1 and T2 serve to achieve the main objective 2 in the catchment area of the Liao River. In T1, upgrading possibilities for municipal wastewater treatment plants in Shenyang are to be sought in the existing stock in the context of the specific Chinese requirements, whereby the purification line and efficiency are to be optimized by means of process engineering approaches. At T2, new approaches for the combined treatment of municipal and pharmaceutical wastewater are to be developed, which could be applied in the Shenyang region in the future.

In both sub-projects, technology and know-how from Germany will be tested and passed on in China and both sub-projects will be realized by operating a semi-technical MBR pilot plant with additional advanced wastewater treatment stages if required. The importance of the subprojects is to complement the already conducted investigations by Chinese research institutions by the long-term operation of the semi-technical pilot plant and targeted individual and preliminary investigations and to collect meaningful results and experiences in the treatment of municipal and industrial (or pharmaceutical) wastewater with the help of the applied concept under local boundary conditions, which can ultimately contribute to an improvement of the water quality in the catchment area of Liaohe.

4.1.1 New Approaches for the Co-treatment of Municipal and Industrial Wastewater

4.1.2 State of Wastewater Treatment and Upgrading Plan in China and Shenyang

Water is a scarce commodity in China. It is used inefficiently in agriculture, and the discharge of domestic and industrial wastewater and diffuse discharges pollute many of China's water bodies, making them available for only limited use. Due to the increasing pollution of water bodies by a growing population and industry, the demand for better wastewater treatment is also increasing. To address these problems, technical solutions are important, but they will not be sufficient on their own without proper management mechanisms.

4.1.2.1 Administrative Structure and Legal Basis for Wastewater Treatment and Water Protection

Two ministries are responsible for wastewater treatment. The Ministry of Water Resources (MWR, reorganised and renamed as Ministry of Natural Resources as of 03.2018) with its assigned institutes is mainly responsible for the distribution and coordination of water resources and for the protection of water bodies and groundwater. In addition, the Ministry of Environmental Protection (MEP), reorganised and renamed as Ministry of Ecology and Environment as of 03.2018, issues the laws relevant for wastewater treatment and monitors the implementation of environmental policies. The Ministry of Housing and Urban Rural Development (MOHURD), which is responsible for the construction and maintenance of the infrastructure for water supply and waste disposal, is also responsible. One of the greatest challenges in the Chinese water sector are the complex administrative structures. In particular, the overlapping responsibilities of the various competing actors have been identified as a problem. In addition, areas of responsibility that are not covered by regulations are also proving to be an obstacle.

The first framework law for environmental protection in the People's Republic of China was the "Environmental Protection Law of the People's Republic of China". The "Law of the People's Republic of China on the Prevention and Control of Water Pollution" details this law in the field of water management. In terms of implementation, the Implementing Rules on the Law on the Prevention and Control of Water Pollution and the Regulations on Issues concerning experimental collection of urban sewage treatment fee are valid. Further regulations exist at provincial level. Parallel to the laws mentioned, standards must be observed which are issued by the central government/ministry in Beijing, but also by the provincial governments. The relevant state standards for wastewater management are listed in Table 4.1.

The effluent concentrations of municipal sewage treatment plants are prescribed by law GB 18,918–2002, similar to the German law GB 18,918–2002. In contrast to Germany, where a certain concentration in the effluent of the sewage treatment plant may not be exceeded depending on the size class, Chinese sewage treatment plants are divided into three standard classes based on the effluent concentration, irrespective of their size class. Table 4.2 shows the effluent concentrations of the Chinese standard classes (SK) with the effluent concentration of "size class 5" (GK) according to the German Wastewater Ordinance. It can be seen that the highest Chinese standard class IA is considerably stricter than the German one.

The definition of the discharge standard is strongly dependent on the water body to be discharged. Depending on the quality category, immission values for surface waters are similarly specified in GB 3838–2002. If sensitive waters or water resource protection areas are involved, higher discharge standards must be achieved. In many priority regions, the state discharge standard is therefore tightened to varying degrees

Table 4.1 Environmental and quality standards relevant to water and wastewater in the People's Republic of China

Title of the standard	Number
Quality standard surface water	GB 3838—2002
Discharge standard of waste water into municipal sewers	CJ 3082—1999
Emission standard from municipal sewage treatment plants	GB 18918—2002

Table 4.2 Parameters in the inflow of the pilot plant

Parameters	Unit	GK 5 (>6000 kg$_{BSB5}$/d)	SK I A	SK I B	SK II	SK III
COD	mg/l	75	50	60	100	120
BOD$_5$	mg/l	15	10	20	30	60
SS	mg/l	-	10	20	30	50
Animal/vegetable fat	mg/l	-	1	3	5	20
Crude oil or similar	mg/l	-	1	3	5	15
Surface active substances LAS	mg/l	-	0,5	1	2	5
N$_{ges}$	mg/l	13	15	20	-	-
NH$_4$-N	mg/l	10	5	8	25	-
P$_{ges}$ (Commissioning before 31.12.2005)	mg/l	1	1/0,5	1,5/1	3	5
P$_{ges}$ (Commissioning after 31.12.2005)	mg/l	1	0,5	1	3	5
Colouring	-	-	30	30	40	50
pH	-	-	6—9	6—9	6—9	6—9
Faecal colony count	/l	-	1.000	10.000	10.000	-

in order to increase the purification capacity of wastewater treatment plants, reduce the discharged pollution loads and thus improve water quality.

4.1.2.1.1 Institutions, Procedures and Challenges

In 2010, a total of 62 billion m^3 of wastewater was produced in China. By 2015, the volume of wastewater had increased to 73.5 billion m^3 with approximately 20 billion m^3 of industrial wastewater and 53.5 billion m^3 of domestic wastewater MEP [12] and [13].

Similarly, the number of municipal wastewater treatment plants increased from about 2800 in 2010 to over 6000 plants in 2015 according to the annual statistical report of the Ministry of Ecology and Environment, which, with a total capacity of 53.2 billion m^3/d, were able to treat a total of about 9 billion m^3/d of wastewater in 2015, more than in 2010. The treatment capacity corresponds to a connection rate of about 73%. Depending on their capacity, the treatment plants are allocated to the following size categories on a percentage basis, in proportion to the total number Jin et al. [8].

- 9% Small sewage treatment plants with capacities up to 10,000 m^3/d:
- 75% Medium-sized sewage treatment plants with a capacity of 10,000 to 100,000 m^3/d
- 16% Large wastewater treatment plants with a capacity exceeding 100 000 m^3/d

In most cases, similar concentrations of wastewater parameters are present in municipal wastewater as in Germany. Due to leaky sewers and the resulting infiltration of extraneous water, however, very low concentrations of COD and BOD5 in wastewater were found in the greater Shanghai area, Kunming and many regions of northern China Yun [25].

From a technical point of view, most municipal wastewater treatment plants in China currently consist of a mechanical treatment stage, primary treatment, biological treatment stage or secondary treatment. Since the introduction of different minimum standards in 2002, investments in the field of wastewater treatment have increased rapidly. The aim was to reduce the content of nitrogen and phosphorus in wastewater. Especially for industrial wastewater, the AAO treatment process, an effective, relatively inexpensive and easy to implement treatment option, is used.

The Urban Water Environmental Department of CRAES Tian [22] published a China-wide overview of 828 wastewater treatment plants with a total wastewater treatment capacity of about 63 million m^3/day (Fig. 4.2). Of the wastewater treatment plants, 2% have only a basic treatment stage, 89% have a mechanical–biological treatment stage and 9% have an additional more advanced treatment stage. The biological stages use the AAO and AO process as the most common technique, followed by aeration ditches and SBR systems. As further techniques the main biofilm processes are used. Process combinations such as SBR, ABR, UASB and CASS are also used to improve the purification performance. All larger (from 100,000 m^3/d) sewage treatment plants have a disinfection stage (UV radiation or chlorination) installed in the effluent to disinfect the treated wastewater.

Despite the existing modern wastewater treatment processes, the actual purification performance of many wastewater treatment plants in China is unsatisfactory. Several reasons for this have been identified:

- complex wastewater composition and unclear discharge source
- low operating temperature in winter
- incomplete monitoring and sampling
- backlogged procedures or investment and energy restrictions

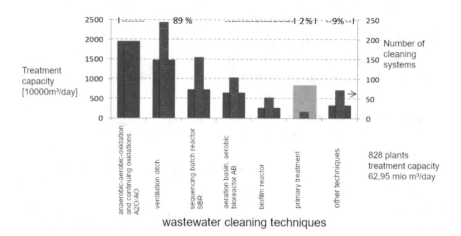

Fig. 4.2 Treatment techniques

- unsafe sludge disposal
- unqualified wastewater specialists and poor operation

For this reason, the Chinese experts attach great importance to upgrading the systems, which is also the aim of T1. Some of the sewage treatment plants are located in densely populated or rapidly developing areas, so that only small areas are available for a sewage treatment plant expansion. Necessary increases in capacity or more extensive purification services therefore generally require an area-neutral optimisation of the sewage treatment plants.

Apart from a few exceptions, most wastewater treatment plants in China use simultaneous aerobic sludge stabilization during biological wastewater treatment. Energy optimization measures such as power generation by sludge digestion are difficult to realize due to several limitations. There are also no plans for the elimination of micropollutants discussed in Germany. In addition, the focus of wastewater disposal in China has so far been on wastewater treatment. This means, however, especially in the case of heavy rainfall events, considerable material pollution of the surface waters through wastewater discharges or wastewater relief. Checking the condition of sewers and optimising the upstream sewerage system are therefore also possible upgrading measures in China Yun [25].

4.1.2.1.2 Environmental Problem and State of Wastewater Treatment in Shenyang

As one of the two study areas, Shenyang is located in the northeast of the People's Republic of China and is the capital of Liaoning Province. Benefiting from government support, Shenyang is an economic and cultural center in the northeast of the country.

Shenyang is located in the floodplain of the Liao and Hun rivers on the southern edge of the northeast China plain. The terrain slopes gently from northeast to southwest. Shenyang has a continental climate characterized by hot, humid summers due to the monsoon and dry, cold and extremely long winters due to the Siberian high pressure system. The average annual temperature in Shenyang is 8.3 °C. The average temperatures are 25 °C in July at the warmest time and about −15 °C in January at the coldest time. Extreme temperatures range from below −30 °C Meteoblue AG [14]. In January 2018, the SINOWATER pilot plant, which was located on the site of the reference wastewater treatment plant in Shenyang, experienced a daytime low temperature of −33 °C. About 500–600 mm of precipitation falls on average within one year and most of the annual precipitation occurs from June to September.

The city of Shenyang has a total area of about 12,860 km^2 and is composed of ten urban districts, two counties and one independent city. The population in 2015 was about 7,300,000 in the metropolitan region. The average population density of the Shenyang metropolitan region is 568 inhabitants/km^2.

The city is known as the "Ruhr Area of the Orient", and since the beginning of the twentieth century has been the economic centre for the whole of north-eastern China and the headquarters of the Chinese manufacturing industry. Similar to its German counterpart, Shenyang has a long history as an important industrial center in northeast

China and is today a location for the automotive, mechanical engineering, chemical, pharmaceutical and financial industries. Due to various problems concerning the technical state of the art, the organizational structure, the environmental pollution as well as the unfavorable climatic conditions of urban design, the whole region of Northeast China is subject to a slowly decreasing economic growth at the beginning of the Chinese reform and opening policy in 1978. In the 1980s and 1990s, many reconstruction measures were therefore carried out in the old Tiexi industrial district in Shenyang, which is one of the largest and most heavily used urban industrial areas in China. In 2002, a new land use plan for the city of Shenyang was presented, according to which many industrial factories were to be moved from the urban core area to new industrial and economic development zones further west Zhang [26]. This relocation was part of a new urban design to accommodate economic shifts in Shenyang. In view of various factors, Shenyang has become increasingly attractive since the 1990s and is attracting the interest of global corporations as well as German small and medium-sized enterprises. One of the city's most advanced projects in the new industrial zones today is the Chinese-German "Intelligent Equipment Manufacturing Industrial Park", another example of the regional administration's efforts not only to promote Shenyang in its role as the world's largest center of machine tool manufacturing, but also to encourage international companies to set up operations in the vicinity of their customers.

In Shenyang, natural and industrial influences result in special requirements for the protection of regional water resources, especially the Liao River in Liaoning Province, which is the largest river in northeast China with a length of 1390 km. As a conventional heavy industrial city in the catchment area of the Liao River, the city of Shenyang has long been seriously confronted with environmental problems, especially the issue of wastewater treatment. The already explained fact that almost 80% of the annual precipitation falls between June and September means that for many months very little runoff is discharged into the surface waters of the catchment area. Especially in the densely populated metropolitan region of Shenyang with over 7 million inhabitants and the numerous industrial enterprises, the large wastewater pipes cause high concentrations of pollutants in the surface waters there. In the catchment area of the Liao River, this particularly affects the tributary Hun and its tributaries Xi, Pu, Baitapu and Mantung.

The Hun River flowing through Shenyang is highly polluted by industry. For example, in 2013, NH4-N and TP concentrations exceeded the minimum requirements of the highest water quality class in China in all tributaries of the Hun River. The high level of pollution can be explained by the many industrial plants, the large population and the low water volume. According to CRAES, pharmaceutical wastewater and agricultural wastewater are the main sources of NH4-N pollution.

The first sewage treatment plant Beibu was built in Shenyang in 1994, and between 2002 and 2007 the first significant expansion of the capacity of the sewage treatment plant took place. Towards 2010, the last large treatment plant and several small treatment plants were built along the Pu River, so that almost all the wastewater from the city districts can be treated and the connection rate has now reached 95%. Since then, industrial wastewater is completely discharged into the sewage treatment plants Shenyang EPB [21].

Figure 4.3 shows the development of the treatment volume of the municipal wastewater treatment plants in the Shenyang urban region since 2004. 2016 about 731 million m^3 of wastewater were treated in Shenyang, i.e. about 2 million m^3/d, which was three and a half times the amount of 2004 Hu [6] and Shan [20].

There are currently 37 wastewater treatment plants with a total capacity of 2.8 million m^3 of wastewater per day: Shan [20].

- 7 sewage treatment plants with >100.000 m^3/d
- 6 sewage treatment plants with 50.000–100.000 m^3/d
- 25 sewage treatment plants with <50.000 m^3/d

It should be noted that municipal wastewater treatment plants are mainly located in the core area of the city. In contrast, the more rural districts together have only a few treatment plants.

Altogether, 83.2% of the wastewater generated in Shenyang is treated, while the average capacity utilisation of the treatment plants is 71.4%. This means that about 405,000 tons of wastewater (presumably industrial and municipal wastewater as well as outdoor discharges) are discharged untreated into the waters in and around Shenyang every day. According to official requirements of the local environmental authority, 28 of the 37 sewage treatment plants have to comply with the minimum effluent concentrations of SC I from the time of commissioning, while the effluent values of the other 9 plants should correspond to SC II. Table 4.3 gives an overview of the number and capacity of the treatment plants depending on the standard classes to be met in Shenyang.

It is clear from this that the 21 treatment plants, which are supposed to comply with effluent concentrations of SK IA, treat only 0.79 million m^3 in total and are not sufficiently utilised. The reason for this is that some of them have not yet been commissioned or are currently undergoing conversion/expansion with the purpose of upgrading. In category SK II there are many large treatment plants with a total

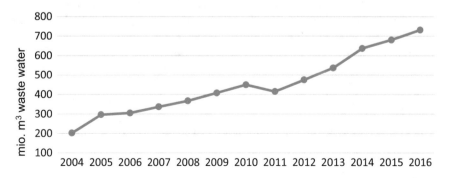

Fig. 4.3 Development of the annual treated wastewater volume in the urban region of Shenyang

Table 4.3 Distribution of municipal wastewater treatment plants and treatment capacity in Shenyang according to Shan [20]

	SK I A	SK I B	SK II
Quantity	21	7	9
Design capacity	1,44 Mio. m^3	0,17 Mio. m^3	1,25 Mio. m^3
actually treated waste water quantity in 2016	0,79 Mio. m^3	0,09 Mio. m^3	1,12 Mio. m^3

throughput of 1.12 million m3 of wastewater, which corresponds to more than 50% of the total wastewater volume Shan [20].

Therefore, unfavourable inlet compositions and missing requirements for process selection and operation are responsible for the effluent concentration of SK II. This leads to an insufficient elimination of pollutants in the wastewater treatment plants. An unpublished study report by the Shenyang Academy of Environmental Sciences shows that the effluent values of nutrient parameters such as TN, NH4-N and TP exceed the limits of the respective classes in many wastewater treatment plants.

Starting in 2018, the environmental authority has set the effluent concentrations of SK IA as binding for the sewage treatment plants in Shenyang. There is therefore a great need for upgrading the old treatment plants, which account for over 50% of the total treatment capacity in Shenyang and cannot meet modern discharge standard requirements in the long term due to the high inflow values. In addition, the city is also making efforts to develop new management measures and technical approaches for the treatment of specific industrial wastewater streams, such as the co-treatment of pharmaceutical wastewater in municipal wastewater treatment plants. These are also reasons for the implementation of the two SINOWATER subprojects T1 and T2.

4.1.2.1.3 Status of Pharmaceutical wastewater Treatment in China and the Pharmaceutical Company "Northeast Pharmaceutical Group Co, Ltd. in Shenyang

China is regarded as the world's largest manufacturer of raw pharmaceuticals and stands out due to the size of the country and its large population with large quantities of pharmaceuticals consumed. The pharmaceutical industry accounts for 1.7% of the gross domestic product of total industrial production in China. In the manufacture of pharmaceuticals, the consumption of raw materials is between 10 and 200 kg per kilogram of active ingredient produced. This process produces a lot of heavily polluted pharmaceutical wastewater (PA), so that 2% of the total wastewater volume in China is attributable to the pharmaceutical industry(Xiao and Zhang [24] Zeng et al. [27].

In China, PA has a complex composition and contains a large number of organic pollutants in high concentrations. The most important of these are COD, BOD5, NH4-N, SS and dyes (including turbidity caused by toxic ingredients). It is also characterised by strongly varying pH values and high salt contents. The amount and composition of wastewater is subject to strong fluctuations Xiang [23].

The circumstances described above call for an improvement or targeted expansion of wastewater treatment systems in Chinese industry. Some of the procedures already applied in the Chinese pharmaceutical industry are described below: Li and Li [11].

Physical–chemical treatment processes:

- Coagulation as the most economical sedimentation process, which not only reduces the concentration of pollutants in the wastewater, but also increases biodegradability.
- Flotation: impurities adhere to gas bubbles, which rise to the surface and can be separated.
- Adsorption processes in which drug residues, but also the COD, the colour and the odour are adsorbed by adsorption effects on e.g. coal.
- Fenton method: In acidic medium, organic substrates are catalysed with hydrogen peroxide by iron salts. In the wastewater treatment plant of a pharmaceutical company in Wuhan, a COD elimination level of over 90% can be achieved by using this process.

Biological treatment processes:

- Aerobic processes: Various aerobic processes are used in the wastewater treatment plants of Chinese pharmaceutical companies: Sequencing Batch Reactor (SBR), Cyc-lic Activated Sludge System (CASS), Cyclic Activated Sludge Technology (CAST), Intermittent Cycle Extended Aeration System (ICEAS), Modified Sequencing Batch Reactor (MSBR) etc.
- Anaerobic process: In China, anaerobic fermentation is currently used as the main approach for anaerobic processes, as it offers several advantages: high organic load degradable, low OEL production, simple sewage sludge dewatering, no aeration required, recovery of biogas possible.
- Combination of aerobic and anaerobic processes: In order to exploit the advantages of both processes, they are often combined. The following combinations are often used in the Chinese pharmaceutical industry: Microelectrolysis - anaerobic hydrolysis and acidification process - SBR; pretreatment - Upflow Blanket Filter (UBF) - contact oxidation - Biological Aerated Filter (BAF); hydrolysis and acidification process - Upflow Anaerobic Sludge Blanket (UASB) - SBR. The wastewater treatment plant of the pharmaceutical company "Northeast Pharmaceutical Group Co., Ltd.

Furthermore, there are other strategies to reduce the negative environmental impact of the pharmaceutical industry in China. These are the implementation of new management systems, clean production facilities in terms of water consumption and pollution, deeper quality control, new management strategies in the companies and the government, and the investigation and application of new processes for pre-treatment, biochemical treatment and advanced wastewater treatment of the wastewater generated by the pharmaceutical industry.

In Germany, the topic of the 4th purification stage or elimination of trace substances has been discussed for several years. As in Germany, Chinese waters

are also polluted by the pharmaceutical industry and by the consumption of pharmaceuticals through pharmaceutical trace substances, which have negative effects
on flora and fauna. It is important to identify the pathways of entry and to prevent
the entry. Furthermore, controls must remain strict in order to prevent the illegal
discharge of polluted wastewater. The aim of wastewater treatment in the pharmaceutical industry should be the elimination of typical wastewater parameters as well
as the elimination of trace substances.

As mentioned above, the SINOWATER pilot plant for sub-project T1 with municipal wastewater for T2 was fed with a mixture of municipal and pharmaceutical
wastewater. The PA is the water from the effluent of the wastewater treatment plant
of the pharmaceutical company "Northeast Pharmaceutical Group Co. Ltd. (NEPG)",
founded in 1946, is one of the largest companies in the field of chemical synthesis
and bio-logical fermentation. In addition, the company's business activities include
Western medicine preparation and micro-ecological preparation as well as the distribution of pharmaceutical products. NEPG produces 12 large series of active ingredients including vitamin series, antibiotics, anti-AIDS drugs, digestive drugs, narcotics
and cardiovascular and cerebrovascular drugs. The company owns the largest production line for vitamin C in the world and the independent technical property right for
purely chemical synthesis of berberine. The company also has a processing base for
feed additives and veterinary drugs based on pharmaceutical raw materials NEPG
[17].

Information on the NEPG's treatment plant and the quality of the wastewater on
site was obtained during visits to the pharmaceutical plant and discussions with the
contact persons. Figure 4.4 shows a diagram of the company's wastewater treatment
plant, which is located on the NEPG site in Shenyang. The two partial water flows
A and B together have a volume flow of 6,000 to 7,000 m^3/day. Due to the high
inflow values of the heavily polluted wastewater, this partial flow passes through two
complete cleaning cycles. Table 4.4 lists the parameters COD, BOD5, pH and SS of
the two partial flows in the feed.

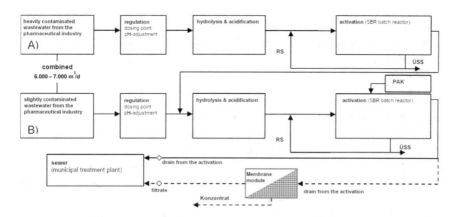

Fig. 4.4 Flow chart of the wastewater treatment plant "Northeast Pharmaceutical Group Co., Ltd

Before the streams are subjected to hydrolysis and acidification, they pass into regulation basins, where various substances can be dosed. The pH value can also be adjusted there. In the anaerobic basin, hydrolysis and acidification take place. The pre-treated wastewater is then transferred to the aeration tanks. The treatment plant has a large number of aeration tanks, as the biological treatment works according to the SBR principle. This means that the sludge is not separated from the purified wastewater in an external secondary clarifier by settling, but the settling process is waited for in the activation tank and the supernatant is then pumped out. There are always basins in different phases of the batch operation, so that a continuous wastewater treatment can take place. The remaining sludge is partly pumped back into the inlet of the activation as return sludge. The rest is discharged in the form of excess sludge.

After the biological treatment of the heavily polluted wastewater, this water undergoes the same purification process - from hydrolysis and acidification - together with partial stream B once again. The aeration tanks in the second purification line of the wastewater treatment plant also operate in batch mode (SBR). In contrast to the first cleaning line, it is possible to dose PAH into the aeration tanks.

In the diagram two possible paths of biologically treated wastewater are shown after the activation tanks of the second treatment line. At the beginning of the project (November 2017 to the beginning of January 2018), the water did not undergo any additional treatment, but was discharged into the municipal sewage system. From mid-January onwards, NEPG put an external membrane stage into operation. A PTFE membrane manufactured in Japan is used in the form of a capillary module, which is constructed from tubular membranes.

The discharge of the NEPG WWTP into municipal sewers is done as described in Chapter 2.1.1 and Table 4.1 explains the administrative structure and legal basis for wastewater treatment and water protection according to the national discharge standard CJ 3082–1999 for indirect dischargers. Officially, however, the treatment plant has to comply with the regional discharge standard DB 21/1627–2008 for Liaoning Province, which was issued in 2008 and is stricter than the state standard. The prescribed minimum discharge values of some main sum parameters are shown in the following table The actual effluent quality of the NEPG treatment plant (as part of the inlet to the SINOWATER pilot plant) during the operation phase for T2 is described in detail in chapter 2.4 due to its complexity.

The pharmaceutical wastewater treated in this way was to be discharged through a separate sewer to a treatment plant, which was newly built in 2016 and has a

Table 4.4 Parameters of the partial water flows A and B of the NEPG wastewater treatment plant

Partial flow	COD (mg/L)	BOD$_5$ (mg/L)	pH	SS (mg/L)
A: heavily polluted wastewater	10.000 – 20.000	3.000 – 8.000	6 – 9	~500
B: slightly polluted wastewater	~1.000	300 – 400	6 – 9	~200

Table 4.5 Minimum requirement of main sum parameters in the national and regional discharge standard into municipal sewer systems (with treatment plant at the end)

Parameters	Unit	national initiation standard CJ 3082—1999	regional induction standard DB 21/1627-2008 (Province Liaoning)
COD	mg/l	500	450/300
BOD$_5$	mg/l	300	250
SS	mg/l	400	300
N$_{ges}$	mg/l	-	50
NH$_4$-N	mg/l	35	30
Phosphates	mg/l	8	5

Table 4.6 Membrane type and TMP of membrane processes Pinnekamp and Friedrich [19]

Membrane processes	pore size	Membrane type	TMP	Operating mode
Microfiltration	0,1 -5 µm	Pore-membrane	0,1 – 3 bar	Crossflow- u. Dead-End-Betrieb
Ultrafiltration	0,005 - 0,1 µm	Pore-membrane	0,5 – 10 bar	Crossflow- u. Dead-End-Betrieb
Nanofiltration	0,001 – 0,01 µm	LD-membrane	2 - 40 bar	Crossflow-Betrieb
Reverse osmosis	< 0,001 µm	LD-membrane	5 – 70 bar	Crossflow-Betrieb

daily treatment capacity of approx. 250,000 m^3 of wastewater, consisting of approx. 7,000 m^3 of pharmaceutical wastewater and 180,000 municipal wastewater. Due to the possibility of treating both wastewater streams simultaneously, this treatment plant was selected as the site for the pilot plant at the beginning of the project. However, due to the lack of a sewerage system and an unresolved operating permit, the treatment plant could only be put into operation from the end of 2017 onwards with highly contaminated industrial wastewater, which caused the relocation of the test plant according to "GWSTP".

4.1.2.1.4 Parallel Research Activities

According to experience, the pre-treated pharmaceutical wastewater as well as the industrial wastewater in Shenyang is generally characterized by poor biodegradability (e.g. unfavorable B/C or C/N ratio), which is a great challenge for the downstream treatment plant with high requirements on effluent concentrations. Therefore, in the past years in Shenyang different considerations for further treatment measures were made and corresponding laboratory tests were carried out. These concerned.

- Adsorption of the hardly degradable components of pharmaceutical wastewater to the biological sludge of a municipal sewage treatment plant
- the ozonation of the pharmaceutical wastewater to break down the refractory substances and
- the hydrolysis and pre-acidification of pharmaceutical wastewater to increase the degradability

Table 4.7 Division of the total project period on site into operational phases

Period		Part-project	Operating phase/ Test series	Purposes
2017	week 19-21	T1+T2	Structure and Commissioning	-
	week 22-26	-	Test phase	Test operation phase with highly polluted industrial wastewater as inflow due to local restrictions
	week 29-30	T1+T2	Relocation and New structure	-
	week 31-39	T1+T2	Run-in phase MBR	Concentration of the activated sludge and adjustment of the inlet to a sufficient return ratio within the plant
	week 40-44	T1	MBR (+GAK)[1]	Control and documentation of the cleaning performance, the membrane and sludge parameters, as well as the discharge after the GAK filter
	week 45-47		MBR + C-Quelle (+GAK)	Test phase to improve the cleaning performance of the test plant by adding an external C source
	week 47-49		MBR + GAK	Adjustment of the feed of PA until the desired mixing ratio of 28% is achieved
2017/ 2018	week 50; 4	T2	MBR + Postozoning + GAK	Test phase for the investigation of the cleaning line of different combinations of the cleaning stages
	week 51; 5		Preozoning + MBR + GAK	
	week 52 — 2		FM/FHM + MBR (+ PAK)[2] (+ GAK)	
2018	week 6-7		MBR + PAK (+ GAK)	
	week 3	T1	FM + MBR	Supplementary test phase for T1 due to an operational failure of NEPG
	week 8	T1	MBR + PAK (+GAK)	Additional test phase for T1 due to the breakdown of NEPG
	week 9 - 10	T2	MBR (+GAK)	Additional test series with increased PA content of up to 100%
	week 11-12	T1+T2	Decommissioning	-

[1] The filtrate must always pass through the activated carbon filter before being discharged
[2] Passive dosing of PAH due to the massively changed feed quality

The combination of sludge adsorption and ozonation proved to be most promising. Figure 4.5 shows the first positive results of the investigations. The BOD5/COD ratio could also be increased from 0.1 to about 0.3 by ozonation. In 2015, a corresponding semi-technical pilot plant was operated by the Chinese cooperation partner at a municipal wastewater treatment plant in Shenyang, where the process combination of sludge adsorption and ozonation is seen as a pretreatment step prior to co-processing in a municipal modified AAO reactor (Fig. 4.6). However, the investigations have shown that with the desired quantity ratio of pharmaceutical wastewater (approx. 30%), the required COD effluent values of standard class IA of 50 mg/l could not be met despite the pretreatment combination (Fig. 4.7) Xiang [23].

Discussions with the Chinese side revealed that no studies have been conducted to date on adsorption on activated carbon. Thus, the process engineering solution envisaged in subproject T2 was developed, which is also regarded by the Chinese side as a promising alternative to the previous solution options. The effect of the

Table 4.8 Overview of sampling points and sampling

Extraction point	Type of sample	Statement	Parameters
Reservoir of GWSTP	composite sample	Feed concentration KA	COD, BOD5, NH4-N, TN, TP, SS
Tanker truck/ Reservoir	sample	Feed concentration PA	COD, BOD5, NH4-N, TN, TP, NO2-N, NO3-N, SS
Ball valve of the coarse material cell	sample	Control of operating sequence	COD
Ball valve of the nitrification tank	sample	Waste water and sludge parameters	SV30, TS, (COD)
Ball valve of the filter chamber	sample	sludge parameters r	SV30, TS
Ball valve of the filtrate pump	composite sample	Outlet concentration MBR	COD, BOD5, NH4-N, TN, TP, SS, (NO2-N, NO3-N)[3]
Ball valve of the filtrate tank	composite sample	Outlet concentration/ Post-ozoning procedure	COD, BOD5, NH4-N, TN, TP, SS
Drain hose of the test facility	sample	Outlet concentration after GAK filte	COD, (TN, TP)

[3]NO2-N and NO3-N were only determined during test phase T2 due to local restrictions

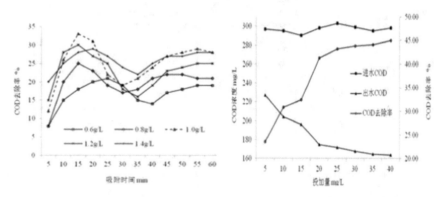

Fig. 4.5 Test results for the adsorption on activated sludge of a municipal wastewater treatment plant (left) and for the ozone oxidation (right) of pharmaceutical wastewater in Shenyang

activated carbon is to be enhanced by the use of a membrane separation stage. In this way, it will be possible to keep the sludge concentration in the activated sludge tank three to four times higher than in other sludge concentrations in order to achieve improved decomposition of the refractory materials (Fig. 4.5).

- Example sludge adsorption: MLSS 1 g/L, T = 15 min., COD Elim. = 33%, BOD/COD = 0,11
- Example ozonisation: Dosiermenge 20 mg/L, T = 30 min., COD-Elim. 0 42%, BOD/COD = 0,31

Fig. 4.6 Chinese modified AAO pilot plant for combined treatment of municipal and pharmaceutical wastewater in Shenyang (FiW)

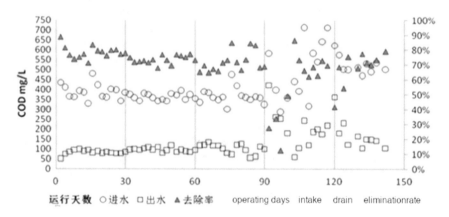

运行天数 ○进水 □出水 ▲去除率   operating days intake drain eliminationrate

Fig. 4.7 COD-related operating results of the pilot plant with 30% pharmaceutical wastewater

4.1.2.2   Innovative Membrane Technology in wastewater Treatment and the Pilot MBR-Treatment Facility in Shenyang

4.1.2.2.1 Membrane Filtration in Wastewater Treatment

Membranes work according to the principle of a filter. They are used for the separation of substances and, depending on the membrane and the process, can separate wastewater components up to a molecular size. The wastewater to be treated is also called "feed" and is separated by the membrane into two phases: the clear effluent as filtrate (or permeate) and the filtered out contaminants as concentrate (or retentate). The morphology is idealised by distinguishing between pore and solution diffusion membranes (LD membrane), so-called dense membranes Baumgarten [1]. Depending on the separation limit (pore size of the membrane), membrane filtration can be further subdivided into microfiltration (MF), ultrafiltration (UF), nanofiltration (NF) and reverse osmosis (RO). The driving force for the mentioned separation processes is the pressure difference between feed and permeate side, the so-called transmembrane pressure difference or transmembrane pressure (TMP). The TMP increases with decreasing separation limit of the membranes used. A summary of the pore size, membrane types or typical TMP and operating mode of the different membrane processes can be found here (Table 4.6).

In municipal wastewater treatment, micro- or ultrafiltration membranes with a pore size of up to 0.1 μm are often used, where almost complete retention of all kinds of bacteria and most viruses could be achieved, while the processes to be applied in industrial wastewater treatment can extend to reverse osmosis due to the requirement to treat highly contaminated wastewater.

In addition to the separation limit, membrane processes can also be distinguished by their design, materials and operating modes. For membranes, a distinction can be made between the two basic forms of flat membranes and tubular membranes. The membranes are arranged in modules. In addition to the membrane, the arrangement and choice of module plays a major role in the cleaning performance of a membrane stage. Since membrane modules are designed depending on the intended use, there is a large number of different module designs. For example, tubular membranes can be used to form tubular modules, capillary modules or hollow fibre modules. Wound modules, cushion modules, disc-tube modules or plate modules are some module forms in which flat membranes are installed Pinnekamp and Friedrich [19].

Both organic and inorganic materials could be used. Organic or polymeric membranes are dominant in all membrane processes (MF, UF, NF, RO) in wastewater treatment. The reasons for this are the lower production costs compared to inorganic membranes and the possibility to select the most suitable polymer for a specific separation problem from the large number of existing materials. Common materials for MF and UF are e.g. polysulfone (PS), polyethersulfone (PES), polypropylene (PP) and polytetrafluoroethylene (PTFE). Inorganic materials, such as ceramics, aluminium and stainless steel, are also mainly used in the MF and UF sector. Advantages over organic membranes are mainly in the mechanical, chemical and thermal resistance Baumgarten [1].

Basically, the two usual operating modes cross-flow operation and dead-end operation dominate, whereby the difference lies in the angle of incidence of the wastewater flow to be treated to the membrane surface. In cross-flow operation, the membrane is exposed to a transverse flow, whereas in dead-end operation the membrane is exposed to the feed vertically Pinnekamp and Friedrich [19].

The membrane bioreactor process (also known as membrane bioreactor process) is the combination of an aeration tank and a membrane filtration. In this process, the membrane stage replaces the secondary settling tank to separate the biologically purified water (filtrate) from the biomass (concentrate). Here, a distinction can be made between two variants:

- Internal membrane stage: membrane module placed in aeration tanks
- External membrane stage: separately installed membrane module outside the aeration tank

In Fig. 4.8 the two mentioned variants of the membrane bioreactor process are schematically shown. If a membrane stage is connected downstream of a biological treatment stage, the aim is to achieve complete solids retention and an extensive hygienisation of the effluent Pinnekamp and Friedrich [19].

In this process, the activated sludge, including all microorganisms, is completely retained by membranes. During operation, the so-called filter cake grows on the membrane surface, which can lead to the formation of a cover layer and must be removed after a certain time in order to prevent permanent blocking of the membrane pores. In all operating modes, the membranes must therefore be back-flushed with air or water (usually by cross-flow aeration). In addition, the effects mentioned below can also lead to a decrease in filtration performance Baumgarten [1]:

- Biofouling: Biofilm formation on the membrane surface
- Colloidal fouling: Accumulation of colloidal (finely divided) dissolved substances leads to a kind of film or slime on the membrane surface
- Scaling: deposits on the membrane formed by crystallization (precipitation)

In the last decades, the membrane bioreactor process has gained more and more importance in wastewater treatment. The advantages of the membrane bioreactor process result from the higher solids content in the aeration tank and the complete retention of particles, microplastics, bacteria and germs by the membranes (and possibly viruses), so that secondary clarification, sand filtration and UV disinfection are not necessary. An improved elimination of organic trace substances can also be enumerated Pinnekamp and Friedrich [19].

The biological stage of membrane bioreactors is designed according to the sludge age, whereby the design sludge age is in the usual range of conventional activated sludge plants. Thus the volume of the activated sludge tank can be dimensioned according to ATV Worksheet 131 (2000), whereby a higher solids content of about 10 to 15 g/L must be assumed. MBR plants can be combined with the usual processes for carbon and nitrogen elimination and for simultaneous aerobic sludge stabilization without any restrictions Krause et al. [4] DWA-M 22738 [10]. The specific

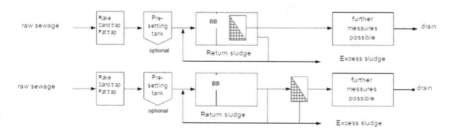

Fig. 4.8 Membrane activated sludge process according to Pinnekamp and Friedrich [19]

excess sludge generation corresponds to that of conventional plants. The municipal membrane bioreactor plants in Germany are currently operated with a sludge age of more than 25 days. The reason for this high sludge age is the lower fouling potential that can be expected Itokawa et al. [7] Judd [9].

The design of the membrane filtration stage depends on the maximum inflow or the area-specific flow for the membrane modules. Decisive design parameters for this are usually the permeability, which for new or cleaned membranes should normally be in the range 150–200 $l/(m^2\text{-h-bar})$ and for intensive cleaning <100 $l/(m^2\text{-h-bar})$, or the membrane flow, which can vary from 5 to a maximum of 25–30 $l/(m^2\text{-h})$ Pinnekamp and Friedrich [19].

In the MBR, which was used during the project, an internal membrane stage in the form of a plate module was installed with a sludge age of more than 20 days in the entire bioreactor.

4.1.2.2.2 Experimental Installation: Design, Construction and Operation

The pilot plant used in Shenyang is a membrane bioreactor with different pre- and post-treatment options. As already mentioned, the entire pilot plant was built as an integrated container wastewater treatment plant (or ship treatment plant) by the German industrial partner MMS in Germany and shipped to the site in China. The process flow is shown schematically in Fig. 4.9 with the construction units. The blue dotted border marks the outer walls of the test plant or container. Dosing equipment and all further cleaning stages are outlined with red dotted lines and were subsequently installed in the test plant on site at the project location in China. The blue turned squares in Fig. 4.9 indicate possible sampling points. The whole system consists of:

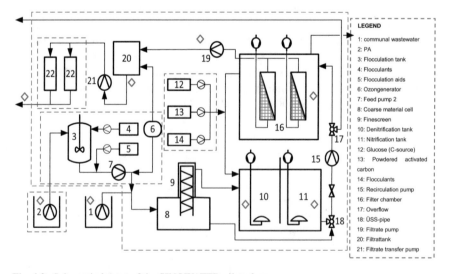

Fig. 4.9 Schematic layout of the SINOWATER pilot plant

- a flocculation plant for the pre-treatment of the raw wastewater,
- a mechanical pre-cleaning unit for the separation of coarse materials with coarse material collection tank,
- an activation tank, consisting of an upstream denitrification and a nitrification tank for biological wastewater treatment,
- a membrane filtration system with immersed flat membrane filtration modules (type: siClaro® FM 622 from MMS AG) to separate the treated wastewater from the activated sludge and
- further treatment stages such as ozonation and activated carbon filter

In order to better understand the local situation of the wastewater treatment plants and the pilot plant, the following figure shows the locations of the GWSTP, the SINOWATER pilot plant and the NEPG together with their in-house wastewater treatment plant. The arrows with the solid lines represent water pipes. The arrow with the green line indicates the main inlet of the pilot plant, and the blue dotted line indicates the PA delivery (Fig. 4.10).

Figure 4.11 shows the test facility on the GWSTP site in summer and winter operation. The reservoir from which the municipal inflow was taken is shown in both pictures. The two storage tanks for the PA, which - like all inlet and outlet hoses of the container - were protected against frost on site with heating cables and foam due to the extreme weather conditions in winter. The pictures on the right show interior views of the plant. Several pictures of the individual operating phases and aggregates can be found in the appendix.

4.1.2.2.3 Project Progress and Operational Phases
The project is subject to different phases. In Germany, the trial period was essentially designed, constructed and transported to the project site on the basis of an evaluation of existing basic principles. After about two months by sea and a successful customs clearance, the pilot plant was delivered to the preselected treatment plant in Shenyang on 09.05.2017.

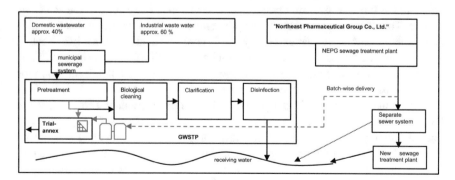

Fig. 4.10 Constellation of the Shenyang municipal wastewater treatment plant (GWSTP) with the pilot plant and the NEPG wastewater treatment plant

Fig. 4.11 SINOWATER experimental plant in summer (top), winter (bottom left) and interior views (right) (FiW)

On site, the piping for the PA inlet and further treatment stages such as the flocculation tank, the filtrate tank for ozonation etc. were further installed. After a frictionless clear water test, inoculation sludge was added to the activation. However, after a 2-month test operation phase, the pilot plant had to be moved to the other GWSTP treatment plant, as it turned out that the new treatment plant built almost 2 years ago could still not be put into operation due to the unclarified state of the sewerage system. The wastewater from the surrounding industrial area, which was delivered in batches as an alternative, was extremely highly contaminated (presumably untreated) and not suitable for commissioning (or sludge inoculation). During the test phase, a large amount of foaming activated sludge was observed from the deaeration pipe despite COD elimination of over 90% after membrane filtration.

The regular operation of the SINOWATER pilot plant extended from 08.2017 to 03.2018. The period can be divided into 3 major operating phases: Construction and commissioning / start-up phase, test phase for T1 with the municipal wastewater generated at the GWSTP treatment plant, and test phase for T2 with co-treatment of pharmaceutical wastewater. Among the last two major test phases, further individual test phases with different purposes are planned. Table 4.7 provides an overview of all test phases.

4.1.2.2.4 Sampling, Measurement Methods and Parameter Determination
The main sampling and sampling points and the parameters determined by them are listed in Table 4.8. To determine the wastewater parameters in the inflow and outflow of the test facility, a mixed sample was taken daily, consisting of three 600 ml samples at the same time. The inflow and outflow samples were taken from the reference treatment plant's receiving tank or from the respective sampling point at 9:00, 12:00 and 15:00 h and mixed in a collecting tank. For the laboratory analysis approx. 500 ml were then filled into a bottle and handed over to the laboratory staff. Furthermore, the sludge parameters were regularly checked by means of random samples from the aeration tank and filter chamber. With the exception of the sampling for the outlet to GAK-Filter all samples could be taken from the cleaning stages by ball valves.

The parameters to be determined for each test phase were selected according to the operating conditions. All analysis results were documented by sampling protocols, which are shown in the appendix.

The MBR test plant was controlled by a switch box with an integrated PLC touch panel. Since the plant was fed with different throughputs, in the test phase T2 with different mixing ratios of municipal and pharmaceutical wastewater, as well as with different additives, the operating parameters had to be constantly changed or adapted. The most frequently changed parameters on the touch panel are:

- Plant throughput
- Aeration time Nitrification (and denitrification)
- Flocculation system settings (including running time of FM and FHM pump, mixing time, total time)
- Volume throughput until the next ÜSS delivery

Determination of the Immediate Parameters
Immediate parameters are the parameters that are measured immediately after sampling. In the project, the following parameters were measured for each in- and outflow mixed sample:

- pH value [-]
- Conductivity [μS/cm]
- Oxygen content [mg/L]
- Temperature [°C]

The measurements of the immediate parameters were carried out with the device "HQ40d" from the company "Hach" (portable 2-channel multimeter). Three different measuring probes were required for the pH value, conductivity and O2 content. The temperature could be determined with the pH probe and the conductivity probe. Furthermore, odour, turbidity and visual colouration of the samples were observed and their changes documented. The on-site parameters give a first impression of the condition and treatability of the wastewater.

Due to the low temperatures during the project duration in winter, measurements with the device were no longer possible and the measuring device had to be moved from the container to the laboratory about 400 m away after a short time. As a result, it was not possible to take oxygen and temperature measurements of the drains (before and after the activated carbon filter) and the inlets after a short time. The measurements of the pH-values and the conductivity could still be carried out in the laboratory of GWSTP.

Determination of the Sludge Parameters

The sludge parameters measured during the tests were:

- TS (Dry matter content) [g/l]
- SV30 (Sludge volume after 30 min) [ml/l]

The SV30 was determined in a 1 L standing cylinder daily for the sludge from the filter chamber and nitrification. Typically, the SV30 should be between 1,000 and 1,500 ml/L during operation. A description of the determination procedure is given in the appendix.

The dry matter content is tested by filtering the sample, then drying the filter residue and weighing it. The dry matter content was also determined during the project for the sludge from the filter chamber and the nitrification. It should be noted that the dry matter content of the nitrification is between 8 and 12 g/L and slightly higher in the filter chamber up to 16 g/L.

The addition of e.g. PAH increases the dry matter content. The SV30 is also affected by this. In order to be able to comply with the mentioned value ranges, the sludge discharge or the plant throughput must be changed or adapted. (MMS, no year).

The following parameters were determined for the pharmaceutical wastewater:

- TR (Dry residue) [%]
- oTR (organic Dry residue) [%]

The difference between the TS content and TR or oTR is that the sample is first filtered when determining the TS content. Thus only undissolved substances are dried. For the determination of the TR the whole sample is dried without previous filtrati-on. To determine the oTR, the dish with the dried sample (after weighing for the TR determination) is heated to about 500 °C, so that all organic residual parts are burnt. The proportions can be calculated from the respective differences.

Determination of the Wastewater Parameters

Table 4.8 already lists the tested wastewater parameters (sum parameters). Daily sampling was carried out in order to check the cleaning performance of the MBR pilot plant and to control the inlet and outlet qualities. The local laboratory of SWSTP could be used during the whole project duration and on behalf of FiW the laboratory staff carried out most of the tests according to the standards of the Chinese Ministry of Environment.

Table 4.9 below lists the determination procedures used for the various parameters according to the respective Chinese standards.

4.1.2.3 Demonstration Results and Recommendations for Upgrading Actions in Shenyang

4.1.2.3.1 Presentation and Comparison of Results

The operation of the pilot plant with the mechanically pre-treated inflow wastewater of the reference wastewater treatment plant GWSTP, which serves the purpose of subproject T1, extended from 01.08.17 to 23.11.17. Furthermore, the two supplemented test series can also be assigned to T1 due to the lack of pharmaceutical wastewater during the T2 phase, namely the tests from 09.01.18 to 14.01.18 and from 19.02.18 to 23.02.18, respectively. The results of the first three series of tests of T1 (see Table 4.7) are presented and analysed in this chapter. For the analysis, calendar data of the sampling were assigned to numbers in the Numbers diagram. This is shown in Annex 4–8 in the Appendix.

Table 4.10 first lists the measured parameters in the inlet of the test plant during operation phase T1. With an average value of 7.9 and a standard deviation of 0.4,

Table 4.9 Determination procedure for wastewater parameters

Parameters	Determination method and standards	Determination limit
COD [mg/l]	Acidified sample is strongly heated with potassium dichromate. From the oxidized potassium, which is equivalent to the oxygen demand, dichromate the parameter. (GB11914-89, 1989)	16 mg/l till 700 mg/l
BOD$_5$ [mg/l]	Sample saturated with oxygen is stored in a closed container under exclusion of light at 20°C for 5 days. The difference between the oxygen content measured before and after is used to calculate the parameter. (HJ 505-2009, 2009)	Undiluted: 2 mg/l till 6 mg/l Diluted: 6mg/l till 6000mg/l
NH$_4$-N [mg/l]	Spectrophotometry of a Neßler reagent to which a part of the sample has been added. Cuvettes and instruments from HACH Lange were used for the determination. (HJ 535-2009, 2009)	from 0,025mg/l depending on the used cuvette
TN [mg/l]	UV spectrophotometry of a cuvette from HACH Lange that has been mixed with part of the sample. Cuvettes and instruments from HACH Lange were used for the determination. (HJ 636-2012, 2012)	from 0,05 mg/l depending on the used cuvette
NO$_2$-N [mg/l]	Spectrophotometry of a cuvette from HACH Lange Cuvettes and instruments from HACH Lange were used for the determination. (Hachlange Schnelltest)	from 0,015 mg/l
NO$_3$-N [mg/l]	Spectrophotometry of a cuvette from HACH Lange Cuvettes and instruments from HACH Lange were used for the determination. (HACH Lange rapid test)	from 0,015 mg/l
TP [mg/l]	Spectrophotometry with ammonium molybdate as oxidizing agent for the decomposition of phosphorus compounds. (GB 11893-89, 1989)	from 0,01 mg/l depending on the used cuvette
SS [mg/l]	Filtering of the sample through filter paper with a pore size of 45 μm and subsequent drying of the filter paper at 103°C to 105°C until the weight is constant. (GB 11901-89, 1989)	-

Table 4.10 Parameters in the inlet of the test plant

	Parameters	n	Ø	Max	Min	s	Unit
Instant Parameters	Water temperature	78	20,9	33,4	9,1	4,4	[°C]
	pH value	82	7,9	8,7	7,3	0,4	[-]
	Conductivity	81	1468,8	1987,0	887,0	199,8	[µS/cm]
	Oxygen content	72	3,3	6,7	0,14	1,2	[mg/l]
Waste water	COD	89	191,4	371,0	102,0	52,5	[mg/l]
	BOD$_5$	65	67,7	143,0	24,1	22,2	[mg/l]
	NH$_4$-N	89	29,6	57,3	14,0	6,8	[mg/l]
	TN	89	34,1	59,6	18,8	6,8	[mg/l]
	TP	55	3,6	12,0	2,0	1,6	[mg/l]
	SS	67	111,1	348,0	24,0	53,6	[mg/l]
Ratio	COD : BOD$_5$	65	3,0:1	6,4:1	1,4:1	1,1	[-]
	BOD$_5$: TN	65	2,1:1	4,7:1	0,9:1	0,8	[-]

the pH value is in the optimum range between 7.5 and 8.5 for a biologically based wastewater treatment. The water temperature is also above the minimum temperature of 12 °C with an average value of 20.9 °C. As can be seen in the table, data sets of different sizes are available for the individual wastewater parameters, especially for BOD5 and TP. The reasons for this were limited capacities, insufficient available materials and defective analysis equipment in the laboratory of the reference sewage treatment plant. Based on the analysis data, an average COD:BOD5 of 3:1 was available in the influent, which is above the optimal ratio for a biodegradability of 2:1. Furthermore, the average COD5:N ratio of 2.1:1 was below the minimum ratio of 2.5:1.

At the beginning of commissioning, the MBR pilot plant first had to be run in. For the time being, the cleaning performance was in the background. Rather, the aim of the start-up phase was to achieve and maintain the following basic requirements for a stable and reliable cleaning performance:

- Inoculation of the aeration tank with activated sludge from the reference treatment plant
- Concentration of the activated sludge to a dry substance value of at least 8 mg/l as well as a basic SV30 value between 1000 ml/l and 1500 ml/l in the aeration tank
- Slowly increase the total throughput to at least 400 l/h and simultaneously
- Prevent the diaphragm pressure from falling below −150 mbar

The MBR + C-Source test series is presented here as an example of the 4 test series with municipal wastewater.

MBR + CSource

In order to investigate whether the cleaning performance of the test plant against nutrient parameters could be improved compared to the normal MBR test series by increased degradability in the influent, glucose was added to the activation (deni zone) as an external C-source in the form of solution by means of a peristaltic pump in the subsequent operating phase. Basic operating data and parameters are shown in Table 4.11. The measured inflow concentrations during the 22-day test phase resulted in a COD:BOD5 ratio of 2.8 and a BOD5:N ratio of 2.2, which is still outside the optimum condition. The aim of the addition is to raise the BOD5:N ratio to about 4.5 to 5. The dosing quantity was based on the optimal requirement for a functioning denitrification on the basis of the worksheet DWA-A 131 and was calculated or adjusted with the average values from the previous concentrations of BOD5 and TN in the feed. The BOD5:N ratio is now an average of 5.5:1, so that sufficient readily degradable carbons should be available for denitrification processes.

However, due to the addition of glucose, sludge production increased, which meant that excess sludge had to be removed from the system more frequently. As a result, the sludge age in this operating phase fell to around 20 days. The high standard deviation of 6.5 days results from the adjustment of the discharge cycle for the excess sludge. Furthermore, the data from the filtration chamber show that the filtration properties deteriorated during this operating phase. The gross throughput initially had to be adjusted occasionally due to the transmembrane pressure being

Table 4.11 Operating conditions and operating parameters of the "MBR + C dosing" operating phase.

	Parameters	n	Ø	s	Unit
Inlet parameters	Net throughput	22	343,3	41,5	[l/h]
	Water temperature	14	15,9	2,6	[°C]
	pH	14	7,6	0,1	[-]
	COD : BOD5 (glucose free)	15	1,8:1 (2,8:1)	0,4 (0,9)	[-]
	BOD5 : TN (glucose free)	15	5,5 (2,2)	1,0 (0,4)	[-]
Aeration tanks	SV30	15	1195	95,8	[ml/l]
	TS	16	10,8	0,8	[g/l]
	Room load (glucose free)	15	0,63 (0,26)	0,06 (0,05)	[kgBSB5/(m³*d)]
	Sludge load (glucose free)	15	0,09 (0,02)	0,01 (0,004)	[kgBSB5/(kgTS*d)]
	Sludge Age	16	20,1	5,0	[d]
	Water temperature Nitrification	15	19,2	0,6	[°C]
	Oxygen content Nitrification	22	1,9	0,2	[mg/l]
	Retention time	22	2,8	0,3	[h]
	Reflux ratio	22	6,8	1,1	[-]
Filtration chamber	Gross throughput	22	429,2	51,9	[l/h]
	Transmembrane pressure	22	-113,6	22,0	[mbar]
	Membrane flow	22	8,6	1,0	[l/(m²*h)]
	Permeability	22	77,1	10,7	[l/(m²*h*Bar)]
	Retention time	22	3,5	0,4	[h]

undershot and therefore averaged only −113.59 mbar. As a result, the membrane flow also dropped to 8.58 l/(m² * h). As a result of the lower membrane flow and the higher average transmembrane pressure, the permeability of 77.05 l/(m² * h * bar) was lower than in the "MBR" operating phase. The reduced membrane performance is due to a stronger top layer formation caused by the higher sludge production, which is triggered by the addition of glucose. As a result of the reduced membrane performance, the net throughput decreased, which slightly increased the return ratio during the operating phase. The average residence time in the aeration tank was about 2.8 h and in the filter chamber about 3.5 h. The sludge parameters and the oxygen content fluctuated within the assumed values.

The calculated increase in COD concentration is shown in Fig. 4.12. As in the operating phase "MBR", there were strong fluctuations in the COD concentration in the feed, which ranged between 110 and 370 mg/l. In order to accustom the microorganisms to the carbon load by glucose, only half of the calculated amount of glucose was dosed by day 94. From day 95 onwards, the COD concentration was increased by 136 mg/l on average. The outflow concentration reached a maximum of 40 mg/l and averaged 25 mg/l. The effluent concentration of 50 mg/l of SK IA was maintained on each measured day. An average COD elimination rate of 92% was achieved, which ranged between 86 and 97% over the period. Not shown are effluent concentrations of BOD5 which do not exceed a concentration of 1.6 mg/l and thus elimination rates between 98% and approximately 100% were achieved.

Figure 4.13 shows the measured nitrogen concentrations during glucose dosing. The NH4-N concentration was completely eliminated so that no concentration could be detected in the vicinity. According to the measured values, a change in the TN concentration in the effluent and the TN elimination rate due to the provision of glucose only occurred slowly from day 95 onwards. The inflow and outflow values measured from then on on 13 days resulted in an average elimination rate of 47%. On day 103 this reached its highest value of 63% and led to the lowest effluent concentration of 10 mg/l. Due to the reduction of NO3-N to elemental nitrogen,

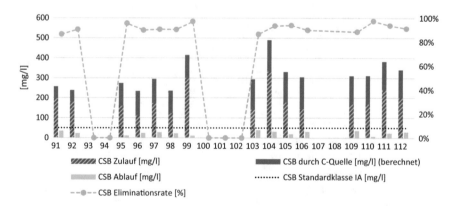

Fig. 4.12 COD concentrations in the inflow and effluent of the pilot plant and elimination rates during the addition of glucose

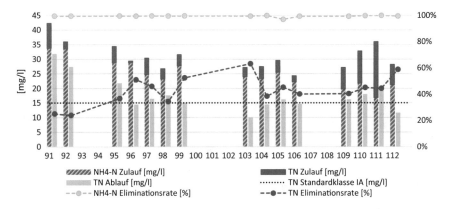

Fig. 4.13 Nitrogen concentrations in the inflow and effluent of the pilot plant and elimination rates during the addition of glucose

which escapes as a gas, an average effluent concentration of 15.37 mg/l occurred. The future required effluent concentration of SK IA (15 mg/l) was met on 6 days.

An improvement in phosphorus elimination through the addition of an external carbon source was also noticeable from day 95. The concentration in the effluent decreased to a value of 0.5 mg/l on day 103, while at the same time the elimination rate reached the maximum value of 82%. The highest effluent value was reached on day 111 with a concentration of 1.4 mg/l. Here, however, the highest concentration was also measured in the inflow of about 4.0 mg/l. Due to the average elimination rate of 71% from day 95 on, the future required effluent concentration of 1.0 mg/l of SK IA was met with an average effluent concentration of 0.90 mg/l. More precisely, the standard was met on 8 of the 12 days measured. The reason for the higher elimination rate is the additional supply of carbon. As already explained, the presence of easily degradable carbon is a basic prerequisite for the incorporation of phosphorus into the biomass-water (Fig. 4.14).

One of the changes in the cleaning performance of suspended solids has not been shown during this operating phase. The average elimination rate of suspended substances was approximately 100%. The following table summarizes the average concentrations and elimination rates of all parameters (Table 4.12).

Comparison of Results

The test results obtained from all test series during the project phase T1 can be evaluated as follows. After successful commissioning, the sludge solid content (dry matter content) in the aeration tank and in the filter chamber was in the range between 10 and 14 g/l, which corresponds to the usual operating values of MBR. The membrane transfer pressure and membrane flows could also be maintained in the desired operating ranges. The throughput of the plant depends on the membrane performance. From the running-in phase to the last operating phase it can be seen that initially the concentration of the activated sludge led to a reduction of the flow or the permeability of the membrane despite the strong and regular air flushing in the filter chamber.

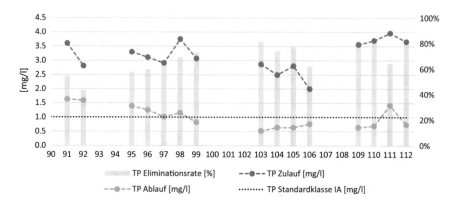

Fig. 4.14 Total phosphorus (TP) concentrations in the inflow and effluent of the pilot plant and elimination rates during the addition of glucose

Table 4.12 Compilation of inflow and effluent concentrations and elimination rates during glucose dosing

Parameters		Inflow [mg/l]		Outflow [mg/l]		Elimination rate [%]	
Description	n	Ø	s	Ø	s	Ø	s
COD	15	315	68	25,0	9,0	92%	3%
BOD$_5$	4	153	18	1,0	0,5	99%	1%
NH$_4$-N	15	25	4,6	0,1	0,2	99%	1%
TN	13	29	3,6	15,4	3,0	46%	8%
TP	13	3,2	0,5	0,9	0,3	71%	8%
SS	14	97	31,0	0,3	0,3	100%	1%

The membrane flow decreased from just under 12 l/(m^2 * h) in the MBR phase to 6 to 8 l/(m^2 * h) in the supplementary phases while maintaining the optimum transmembrane pressure range with the aim of achieving the best possible purification performance in the research project. In practice, a constant throughput is also possible, however, taking into account that the membrane pressure must not exceed the maximum upper limit (in the case of the project −300 mbar) and that the cleaning performance can be impaired by increasing the pressure.

In addition, the performance of the membrane was impaired by the addition of glucose, flocculant and PAH, as the increased sludge production resulted in a stronger top layer. However, membrane filtration allows the adjustment of higher TS contents, resulting in a higher sludge load and a reduction of the reaction volume.

A comparison of the average effluent concentrations and elimination rates of the operating phases assigned to T1 with the literature values of the MBR plants in practice with simulated precipitation is shown in Table 4.13. The average COD effluent concentrations of the MBR pilot plant correspond to the literature values. Deviations between the test series can be found, but this is due to a high and fluctuating proportion of industrial wastewater or inert COD in the inflow. The effluent concentrations of BOD5 are also very low. Easily degradable organic carbon was almost completely

Table 4.13 Comparison of the average effluent concentrations (a, mg/l) and elimination rates (e) of the pilot plant with literature values of MBR plants with simultaneous precipitation (based on DWA [18] Dohmann et al. [2] Pinnekamp et al. [3]

	MBR			MBR + C- Source			MBR + C+ FM			MBR + C + PAK			Literature value MBR plant with simultaneous precipitation
	n	a	e	n	a	e	n	a	e	n	a	e	
COD	22	19,0	90%	15	25,0	92%	5	28,0	86%	5	19,2	88%	< 30
BOD$_5$	11	1,2	98%	4	1,0	99%	-	-	-	-	-	-	<3,0
NH$_4$-N	22	0,1	99%	15	0,1	99%	-	-	-	-	-	-	<2,4
TN	22	28,0	16%	13	15,4	46%	5	10,2	68%	5	13,4	53%	< 13
TP	19	1,4	58%	13	0,9	71%	5	0,15	95%	5	0,11	96%	< 0,3
SS	22	1,6	99%	14	0	100%	-	-	-	-	-	-	0
TS	22	12,1	-	16	10,8	-	5	16,7	-	5	16,8	-	8 - 20

eliminated, which is why a very low concentration in the effluent was detected. This results in a good cleaning performance of the pilot plant of carbon compounds in the wastewater. The addition of flocculants and PAHs could not massively improve the COD effluent values. The COD concentration could be further reduced by a downstream activated carbon filter. However, this measure is not very interesting for upgrading municipal wastewater treatment plants because of the already very good effluent values in the filtrate of the MBR (far below the future limit value of 50 mg/l).

With regard to the effluent concentrations of nitrogen compounds, the literature values of the EN of MBR plants were not complied with in test series of the "MBR" without further measures. Almost no NH4-N was detected in the effluent, which can be justified by the complete oxidation to NO3-N. In this case, the converted NO3-N was not reduced to elemental nitrogen, so that the effluent concentration of 28 mg/l was more than twice the expected concentration of less than 13 mg/l. Responsible for the incomplete denitrification may be the oxygen carry-over from the continuously aerated filter chamber. The continuous aeration, which enabled the cross-flow operation of the membrane filtration, could cause an undesired oxygen input into the wastewater-activated sludge mixture, which flowed via the overflow back into the upstream denitrification tank. As a result, the setting of an anoxic environment could be impaired, which could lead to a disturbance of the denitrification processes. Another cause is the bad C-N ratio in the raw sewage, which has already stoichiometrically strongly limited the maximum degree of denitrification. As a result of the addition of a C source, the TN effluent concentration is significantly reduced. Nevertheless, with improved BOD5:N ratios the literature value or the limit value 15 mg/l of SK IA was not continuously reached. It should be noted that the term TN or Nges in the German wastewater Ordinance only covers the inorganic nitrogen fractions. The wastewater produced contains a large proportion of industrial wastewater in which hardly degradable organic nitrogen compounds are present. For this reason, the effluent values of the "MBR + C-Source" operating phase are already comparable with the values in practice. However, the period of 24 days for the operation phase "MBR + C-Source" can also be the reason for an incomplete formation of the anoxic biocoenosis. Due to the limited time frame of the project, the operating phase could not be extended. Although the data density and the reliability of the results are limited, the two supplemented test series during the project phase T2 showed a further improved denitrification performance with lower effluent values, which could confirm the positive effect of glucose addition and its reasonable dosage.

Furthermore, the TP effluent concentrations of the test plant showed relatively worse values than the literature value during the operating phase "MBR". However, this value refers to MBR plants whose TP elimination rates have been improved by simultaneous precipitation. The biological phosphorus elimination was improved by the addition of C-source, whereby the effluent concentration dropped from an average of 1.4 mg/l to 0.9 mg/l and was already below the limit value of 1.0 mg/l. The achieved average TP elimination rates of 58% without C-source and 71% with C-source show that the pilot plant demonstrates biological phosphorus elimination. The addition of a chemical phosphorus elimination in the supplementary test series "MBR + C + FM" caused the effluent concentration to drop rapidly to 0.15 mg/l, which was fully in line with the literature values. Due to the limited time period, the simultaneous precipitation could only be investigated for a few days.

The analysis results of the SS concentrations confirm the desired values. Although concentrations were found in the analysis results in the effluent, these are due to measurement errors or unclean sampling. The complete retention of solids as well as of bacteria and viruses is a clear advantage for the use of the membrane activated sludge process.

For the wastewater produced in Shenyang, the addition of PAHs could not provide a significant cleaning performance. Due to fluctuating elimination rates and limited data capacity, no reliable assessment of the cleaning performance of powdered activated carbon could be made. The dry matter content increases with the addition of FM and/or PAH, but according to literature values it is still within the controllable range (<20 g/l). All in all, the pilot plant delivers an excellent cleaning performance when treating the rather industrial "municipal wastewater".

4.1.2.3.2 Comparison of the Treatment Performance of the Pilot Plant with the Reference Treatment Plant "GWSTP"

In order to be able to evaluate the treatment performance in view of the nature of the wastewater, a comparison between the effluent concentrations of the reference treatment plant "GWSTP" and the pilot plant will be made, including the target effluent concentration, for classification in the SK IA. In this respect, Fig. 4.15 shows a comparison of the treatment performance of two representative test series of the test plant and of the large sewage treatment plant during the test period of project phase T1. First of all, it is noticeable that the test plant showed considerably better effluent concentrations than the reference treatment plant. It should be noted that the average effluent concentrations of the reference treatment plant are much higher, as they were analysed daily by the laboratory.

Apart from the number of values measured, there were also small differences between the sampling from the reference sewage treatment plant and the sampling from the experimental plant (Chapter 2.2.4). Due to the 24-h mixed sampling in the inflow and outflow of the reference treatment plant, the analysis results represent

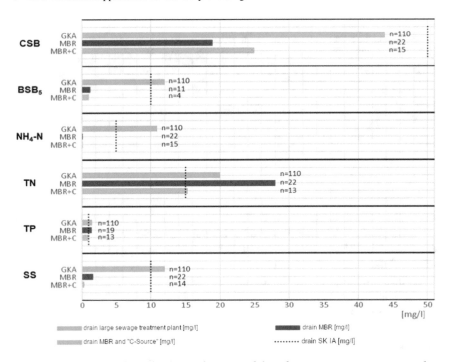

Fig. 4.15 Comparison of the cleaning performance of the reference wastewater treatment plant "GWSTP" with the cleaning performance of the test plant in the operating phases "MBR" and "MBR + C-Source"

the average wastewater quality of the whole day, whereas the three-time sampling in the inflow and outflow of the test plant is only representative for the period from 9 a.m. to 3 p.m. During this period, more polluted wastewater was discharged due to the working hours of the industrial plants. Since the effluent values of the test plant correspond to the literature values and the cleaning performance of the membrane stage is relatively stable, these values can be used for a qualitative comparison with the effluent values of the reference treatment plant.

The COD and BOD and NH4-N effluent concentration of the pilot plant is on average lower than in the reference treatment plant in almost every operating phase. This is due to the fact that the large sewage treatment plant did not have to comply with standard class IA before 2018. As already mentioned, the former plant design and faulty technical management are reasons for the high effluent concentration. In 2005, the large sewage treatment plant was designed with the main tasks of eliminating carbon and ammonium nitrogen and thus comply with the effluent concentrations of standard class II. In order to be able to compare the load conditions of the two plants better, the following Table 4.14 shows the average room and sludge load of the two plants during the project phase T1. As "GWSTP" has additional carrier materials in the aeration tank operated with the HYBAS process, it is difficult to determine the exact sludge load. It is assumed that the use of free-floating carrier materials causes about 10% more biomass in the hybrid activated sludge zone.

Table 4.14 Comparison of the room and sludge load of the reference wastewater treatment plant "GWSTP" and the pilot plant in the operating phases "MBR" and "MBR + C-Source"

	Room load [kgBSB5/(m³*d)]	Sludge load [kgBSB5/(kgTS*d)]
MBR	0,37	0,03
MBR + C-Source	0,63	0,09
Water treatment plant „GWSTP"	0,35	0,12 (0,14*)

*: without consideration of carrier materials

Without the addition of an external C-source, the low room load of the test plant during the normal operation phase "MBR" is very close to that of the wastewater treatment plant, which reflects a good comparability of the two plants. It is noticeable that the low sludge load of the test plant of 0.03 according to DIN EN 12,556–6 rather points towards sludge stabilisation due to the high dry matter content and the guaranteed aeration, whereas the sewage treatment plant with 0.12 is only adjusted to the purification target nitrification, which can explain its limited purification performance in terms of N elimination. The average high TN effluent concentrations of the wastewater treatment plant and the operating phase "MBR" are initially due to the poor BOD5:N ratio in the inflow. The difference between the two effluent concentrations can be assumed to be due to the additional coarse material cell in the test plant, which has removed some of the easily degradable carbon required, in addition to the reasons already clarified (oxygen carry-over). Furthermore, inadequate aeration of the sewage treatment plant may have led to anoxic conditions in the basin, which favoured certain formation of denitrification processes. With the addition of C-source, the sludge load of the pilot plant could be well adjusted to the target range of N-elimination and lower effluent values could be achieved.

The experimental plant did not show a significantly better performance in P-elimination without further measures (C-dosage or simultaneous precipitation). As expected, the SS concentration in the effluent of the treatment plant is significantly higher than the effluent concentration of the test plant. In summary, the test results show that the treatment performance of the test plant is better than that of the reference treatment plant without consideration of specific investment and energy consumption and leads to much lower concentrations of the wastewater parameters in the effluent. This treatment performance can be improved, especially with regard to nitrogen and phosphorus compounds, by adding a C-source and flocculants, since the nature of the wastewater to be treated alone does not allow for a better biological treatment.

4.1.2.3.3 Recommendations for Upgrading the Old Municipal Wastewater Treatment Plants in Shenyang

In principle, the investigations show that the pilot plant has a better purification performance than the reference sewage treatment plant in Shenyang. In view of the poor elimination rates of BOD5 and NH4-N, however, this is not due to the technology used, but to poor management of the reference treatment plant. This leads to the fact that due to insufficient aeration and circulation of the aeration tanks, the optimal conditions for the biological carbon degradation and a functioning nitrification are not created. Therefore, the requirements for the technical operation should be adapted so that the purification performance of the process used can be improved.

On the other hand, some large municipal wastewater treatment plants in Shenyang do not aim for extensive nutrient elimination (cf. Chapter 2.3.2) due to the less stringent environmental requirements (required emission standard) in the year of construction, which is why the "GWSTP" lacks a separate denitrification zone and direct comparability with the pilot plant is limited. As a result of the tightening of the emission standard in the 13th five-year plan (from II or IB to IA) by the environmental authority, upgrading measures are mandatory at the relevant wastewater treatment plants in Shenyang from 2018. The GWSTP is primarily the conversion of the existing old biological stage into AAO basins, which are operated in hybrid mode with floating carrier materials and should provide a better cleaning performance with respect to nitrogen and phosphorus..

Furthermore, the C:N ratio in the inlet should be adjusted to allow denitrification processes to take place. The change in the C:N ratio can be achieved by reducing the amount of extraneous water and rainwater in the inflow, which is of great importance for Sponge City and for the effective purification performance of the wastewater treatment plants in Shenyang. This will lead to higher concentrations of wastewater parameters in the influent. The question arises whether the sewerage system should be regularly checked for leaks and damage and repaired if necessary to avoid dilution of the wastewater by extraneous water.

The additional discharge of industrial wastewater also results in a high load of easily degradable carbon. Furthermore, the dosing of an external carbon source can contribute to an improvement. For the conversion there is the possibility to control the substrate dosage into the aeration tank by means of a nitrate measurement in order to adjust the optimal C:N. The proportion of biologically non-eliminable organic residual N in the influent could be determined, for example, by the results of the test plant (maximum achievable TN effluent values and elimination rate under C overdosage or empirical values).

The biggest challenge for the GWSTP treatment plant are the industrial waste waters to be treated. This applies all the more, because in the future still further industrial enterprises are to be attached to the purification plant and clear production increases of the enterprises with accordingly increased wastewater quantities are expected. For this reason, a complete indirect discharge register must be drawn up, which should include all important and up-to-date information on the plants and their discharge status and is currently only available in small quantities. Due to the fact that many sewage treatment plants in Shenyang have only very limited access to it, the local environmental authorities responsible must exert influence here. The connection situation of the sewerage system must be examined more closely in order to be able to make more precise statements about the discharged industrial wastewater streams. On the other hand, online monitoring of the pre-treatment plant processes is urgently required, at least for the main companies. In order to be able to detect possible faults in the downstream sewage treatment plant at an early stage and to initiate appropriate measures such as commissioning the emergency basin or dosing powdered activated carbon, the online data must be transmitted directly to the sewage treatment plant.

From the point of view of cleaning technology, a changeover to the membrane bio-reactor process would only make sense in the short term if the cleaning performance of the Shenyang treatment plant in question is not sufficiently improved despite optimised technical management, conversion measures and wastewater conditions. Due to the usually low inflow concentrations, it was assumed in the conceptual planning that the use of MBR for the GWSTP treatment plant is not necessarily immediately sensible.

According to the already published annual plans or the current environmental policy of the Chinese central government, even stricter discharge standards will be required in the future, so that the upgrading of the old sewage treatment plants can take place. The definition of the discharge standard depends strongly on the water to be discharged. If sensitive waters or water resource protection areas are involved, higher standards must be achieved up to surface water quality. For this reason, the state discharge standard IA is tightened to varying degrees in many priority regions such as Beijing and the Taihu catchment area. In Shenyang and the Liaohe catchment area, this trend is also expected to continue in the near future. Figure 4.16: Comparison of the maximum achievable effluent quality of the pilot plant with the standard of the Chinese emission and surface water qualities shows the maximum achievable effluent quality (average values of test series) of the pilot plant and the standards of the Chinese emission and surface water qualities. With the exception of TN, most of the achievable effluent values comply with the Chinese standard of surface water quality class III, which is suitable for use as fishing water or for drinking water treatment. In this sense, the use of MBR in Shenyang or the catchment area of Liaohe can be advocated in the long term.

In view of higher operating costs due to higher energy consumption, it must be weighed up whether the advantages of MBR contribute to a meaningful improvement in cleaning performance in each individual case. From the results of the trials carried out in the project, the Sinowater concept offers the following advantages for upgrading the wastewater treatment plants in Shenyang:

- The choice of a sustainable German system with low-maintenance flat membranes (as in the pilot plant) guarantees a minimum operating time of 10 years for the first time.
- The process option tested in project phase T1 results in a higher biomass concentration (10–12 g/l), a desirable higher sludge age and thus a considerable saving of reactor volume and space (4 times as much as GWSTP with a TS content of 2.5 g/l), which can be of great importance for the rapidly developing Chinese metropolises like Shenyang.
- The flow paths of the wastewater in the sewage treatment plant area are thus simplified.
- A germ-free and solid-free effluent is created which enables, for example, a reuse of the wastewater.

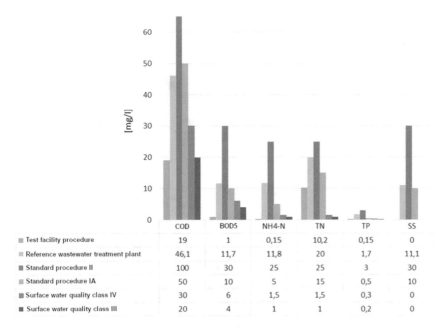

	COD	BOD5	NH4-N	TN	TP	SS
▨ Test facility procedure	19	1	0,15	10,2	0,15	0
▨ Reference wastewater treatment plant	46,1	11,7	11,8	20	1,7	11,1
■ Standard procedure II	100	30	25	25	3	30
▨ Standard procedure IA	50	10	5	15	0,5	10
■ Surface water quality class IV	30	6	1,5	1,5	0,3	0
■ Surface water quality class III	20	4	1	1	0,2	0

Fig. 4.16 Comparison of the maximum achievable effluent quality of the pilot plant with the standard of Chinese emission and surface water quality

- Secondary settling tanks can also be dispensed with. Necessary further treatments, which have to be requested by the environmental authorities at short notice or in the future, such as odour removal by means of a soil filter, can be carried out in existing saved plant components.
- Simultaneous dosing of external carbon sources into the MBR can be easily controlled. In case of disturbances caused by industrial wastewater, the dosing of powdered activated carbon or flocculant into the MBR is recommended.

In the case of upgrading by the MBR process, Fig. 4.17 shows 3 implementation variants based on the recommendations of the DWA, for example at the reference wastewater treatment plant GWSTP, with the concept of not building the nitrification stage separately but integrating it into the membrane stage. In the first option, the secondary clarifier is to be converted to an upstream denitrification stage and the aeration tank is to be used as a membrane stage, whereas in the second option the circuit is to be reversed. The tank volumes saved in this way can be used for another purpose. In the third option, the secondary settling tank can be completely converted for further treatment to improve the cleaning performance. Depending on the required dimensions, the conversion measure should be carried out taking into account the existing tank volumes. For example, a ratio of 1:1 is recommended for the volume ratio of denitrification and nitrification tanks in order to reduce oxygen carry-over through the return flow. In general, calming zones before the sludge recirculation should be provided in practice to avoid oxygen carry-over into the denitrification

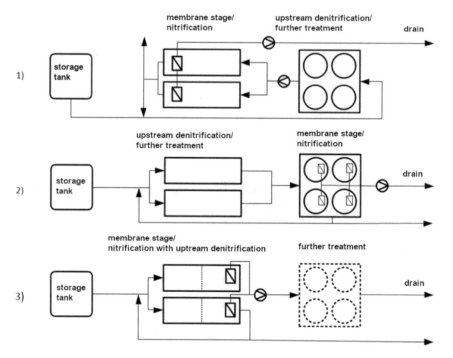

Fig. 4.17 Proposals for upgrading the reference wastewater treatment plant GWSTP to the MBR process

area or to optimise oxygen utilisation in the nitrification area. In addition, in the case of the spatially separated filtration area, it is also advisable to return the return sludge to the nitrification zone. In this way, both sludge circuits can be set separately from each other Pinnekamp and Friedrich [19]. In order to maintain the performance of the membrane modules in the long term, the mechanical pre-treatment would have to be supplemented with a grease trap which can be integrated into the aerated grit trap.

4.1.2.4 New Approaches to the Co-treatment of Pharmaceutical Wastewater in the Municipal Treatment Plant

In this chapter the results of the individual test series from the large project phase T2 (co-treatment of pharmaceutical wastewater) are presented first. In order to ensure the comparability of the different test series, not all parameters were examined due to the large amount of data. Due to the fact that parameters such as NH4-N, SS and BOD5 could be almost completely eliminated in all test series, the inflow and outflow values of COD are therefore preferably compared and the elimination rate calculated for each test series. For most of the test series, however, additional measured data of nutrient parameters are given, which allows statements to be made about the effectiveness of the respective procedure.

The project phase began on 24.11.2017, the date on which the pharmaceutical wastewater was finally available after several local clarifications and coordination. The PA proportion in the total inflow is always given as a mass percentage. According to measurements of the local laboratory, the density of the two wastewater substreams corresponds. Like the presentation of project phase T1, each measured day is assigned a number between 1 and 87 (see appendix). Since the series of measurements did not all take place in a coherent manner, the horizontal axes of some diagrams will not show any coherent numerical sequences.

Near the SWSTP, a new treatment plant will be in running-in operation from 10.2018. This treatment plant will also be fed with the mixed inlet of municipal and pharmaceutical wastewater. The expected PA content will be 28%, so that during the project a PA content of 28% was aimed for in order to test the conditions to which the new sewage plant in Shenyang will also be exposed. On this basis, a recommendation for further purification processes will be made, which could be considered for this and other wastewater treatment plants in the region. In order to test the treatment limit of the pilot plant, the PA content was further increased to 100% at the end of the project. The cleaning performance of the pilot plant will be evaluated for the mixed PA and municipal feed.

4.1.2.4.1 Description Of the Inflow Situation of The Pilot Plant

The total inflow consists of the partial flow municipal wastewater and the partial flow Pharma wastewater. The municipal wastewater, which was treated in the test plant, was taken unchanged as before from the inlet to the biological treatment stage of the "GWSTP" above the submersible pump. Apart from the COD values, most of the inflow values of this partial stream could be assumed to be relatively constant during the entire project phase, with only a few larger deviations. The following diagram shows the concentrations of COD (mean value 176.10 mg/l), BOD5 (mean value 68.35 mg/l), TN (mean value 30.66 mg/l), and TP (mean value 3.02 mg/l) as representative values. The proportion of NH4-N in TN is on average 80%.

As explained in Chapter 5.1, the pharmaceutical wastewater (PA), which was also treated in the pilot plant, has already undergone operational pretreatment, was taken from the effluent of the NEPG treatment plant and delivered to the pilot plant in batches as required. The concentrations of the main parameters in the pharmaceutical wastewater are also shown in Fig. 4.18. It is noticeable that the concentrations of the constituents in the PA vary greatly from batch to batch compared to the municipal partial flow, which means that the load on the biology in the test plant also varied greatly. The commissioning of the membrane stage from day 52 in the NEPG wastewater treatment plant improved the effluent values of the treatment plant considerably, especially for the COD and BOD5.

Figure 4.19 shows the proportion of PA and the COD values representing the mixed feed (COD from the two partial streams and from the mixed feed). It can be seen that the proportion of PA mainly had an influence on the COD feed values of the mixed feed when the COD feed concentrations of PA (before day 52 or commissioning of

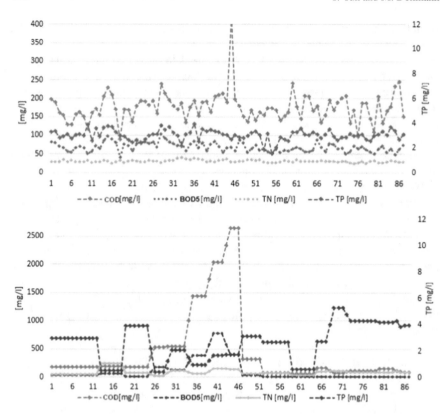

Fig. 4.18 Inflow values of the partial flow municipal wastewater (top) and the partial flow pharmaceutical wastewater delivered in batches (bottom) over the measured period for T2

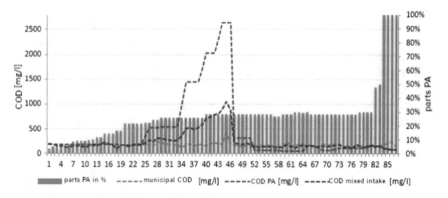

Fig. 4.19 Influence of the PA content on the mixed feed (COD)

the mixed feed) were very high. However, no direct conclusions can be drawn from the level of the PA content itself. The downstream municipal wastewater treatment plant of Shenyang is also confronted with this circumstance, as the discharge values of the operational wastewater treatment plant of the pharmaceutical company (and presumably other companies) are subject to strong fluctuations. It can also be seen that the COD:BOD5 as well as the BOD5:N ratio in the partial streams - and thus also in the mixed feed - is not at all optimal. Especially in PA, the BOD5 values are mostly below the TN concentrations due to the pre-treatment and do not exceed 30 mg/l from day 46. In order to set the desired ratio, the dosing of C-source into the revitalization was continued during the project phase T2. The glucose solution (175 g/L) had a COD value of about 187,000 mg/l.

The individual batch duration (shown in Fig. 4.18) changes during the entire project period, especially from the beginning to day 22 due to the adjustment of the coating ratio until the desired proportion of PA is reached, and from day 81 due to the further increase of the PA proportion up to 100%. A slight fluctuation of the PA content during the stable feeding phase between day 22 and day 81 can be attributed to the occasional operational disturbances and sludge discharge. The working conditions during the entire project period (measuring days 1 to 87), which influence the plant performance, are shown in the table below, together with the standard deviation. Operating conditions and parameters for individual operating phases/series of tests can be found in the appendix (Table 4.15).

4.1.2.4.2 Presentation of the Results of Important Operating Phases
According to Table 4.5 different test series have been carried out. The results of the MBR, MBR + GAK, MBR + ozone + GAK and MBR + PAK test series are presented below as examples.

MBR
In this chapter, only the days are examined on which the pilot plant was tested without the application of further purification stages. A total of 34 measuring days can be assigned to this phase. As already mentioned, two membrane modules of the type FM 622 of the company siClaro® are installed in the filter chamber, so that the membrane surface, the flow rates as well as the purge air requirement of a single module can be doubled. The limit data of a module and the plant are shown in the appendix. The flow and the membrane pressure must not exceed the upper limit ($22 \, l/(m^2 * h)$ or -300 mbar), otherwise the membranes may be damaged. A first automatic error message was given at -150 mbar. According to the manufacturer's specifications, the membrane pressure should be below -100 mbar for optimum cleaning performance, as explained, which was also to be aimed for during the operating phase. The membrane pressure depends on the sludge properties, the wastewater constituents, the throughput of the filtrate pump (in the course of which the membrane flow and permeability) and the flushing air volume flow. Especially by changing the flow rate as well as the dry matter content, the membrane pressure could be significantly influenced, as shown in the diagram below (Fig. 4.20, n = 34). Each measured value for the diaphragm pressure is a representative value of the respective measuring day. In

Table 4.15 Operating conditions of the test facility for project phase T2

	Parameters		n	Ø	s	Unit
Inlet parameters	Net throughput		87	292,3	53,2	[l/h]
	pH		85	7,69	0,8	[-]
	COD : BOD₅ (without glucose)		85	2,3:1 (4,8:1)	2,0 (3,0)	[-]
	BOD₅ : TN (without glucose)		85	4,1 (1,1)	1,4 (1,2)	[-]
Aeration tanks	SV30		64	1053	145	[ml/l]
	TS		64	12,8	1,9	[g/l]
	Room load	BSB	76	0,56	0,25	[kg/(m³*d)]
		CSB	87	1,28	0,64	
	Sludge luad	BSB	64	0,045	0,026	[kg/(kgTS*d)]
		CSB	64	0,098	0,064	
	Sludge age		87	22,9	5,1	[d]
	Water temperature Nitrification		60	19,0	1,0	[°C]
	Oxygen content Nitrification		87	2,0	0,5	[mg/l]
	Retention time		87	3,2	0,5	[h]
	Reflux ratio		87	6,1	2,1	[-]
Filtration chamber	Gross throughput		87	365,4	66,1	[l/h]
	Transmembrane pressure		87	-96,6	22,0	[mbar]
	Diaphragm flow		87	7,5	1,3	[l/(m²*h)]
	Permeability		87	84,7	25,2	[l/(m²*h*Bar)]
	Retention time		87	4,1	0,6	[h]

addition, the membrane pressure is influenced by the addition or residual amount of various substances such as PAH or FM remaining in the system (from day 54, increased DM values as a result of the residual effect of FM or PAH dosing). It can be seen that the diaphragm pressure decreases or increases with a slight time delay in response to a change in flow. The correlation for the displayed curves is R = 0.59. After shifting the measuring days by two days, the correlation is R = 0.65.

The working conditions prevailing during the test phase are comparable with those shown in Fig. 4.21. Figure 4.21 shows the COD inflow and outflow values of the test plant as a function of the PA content (n = 34). The respective COD contents in the mixed feed can also be seen. As already mentioned, glucose was added to the mixed feed in order to support the decomposition of nitrogen. As a reaction to the high TN feed values of the PA (>82 mg/L) of the test phase "Further increase of the PA proportion", the glucose dosage was greatly increased from day 82.

It can be seen that the limit value of 50 mg/l can almost always be adhered to in the beginning. In the rear part of the diagram, however, it is exceeded, which is mainly due to the PA content of the mixed feed, which in the end has been increased to 100% over a few days. It is noticeable that the effluent concentrations, especially from day 54 onwards, are for the most part higher than the PA concentrations, and for the last four days only slightly lower than the PA concentrations, which indicates the extremely

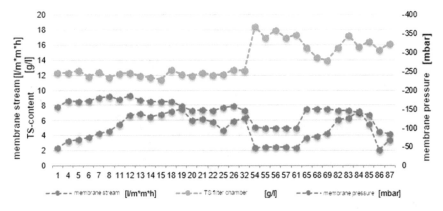

Fig. 4.20 Relationship between membrane flow, dry substance content and membrane pressure

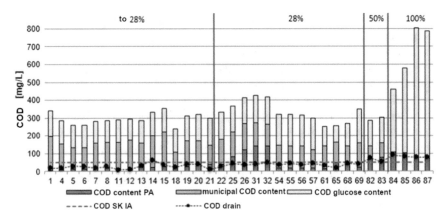

Fig. 4.21 Inflow and effluent concentrations as a function of PA content during the operating phase MBR

poor degradability of the COD in the PA (especially in the PA already subjected to the operational membrane stage, in contrast to the COD content of glucose). In order to indicate this phenomenon, Fig. 4.22 shows the correlation between the COD elimination rate (related to the mixed feed without considering an additional C source) and the PA content in the mixed feed. On the other hand, it can be interpreted from the slidegram that in the industrial "municipal wastewater", a part of the COD can also neither be eliminated by biological stage nor by ultrafiltration. In order to be able to effect a degradation of the inert COD, further purification processes were tested (see following chapters).

During the operating phase, the future TN limit of 15 mg/l can only be complied with on three days with the addition of additional C-source. Despite the fluctuation in the mixed feed, a stable and orderly elimination line can be seen up to day 61. The poor effluent values on the rear days confirm that a large part of the nitrogen present

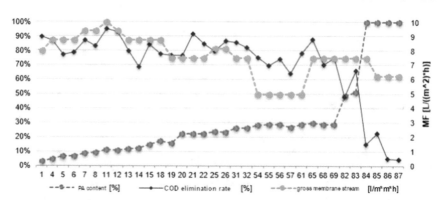

Fig. 4.22 Membrane flow & correlation between PA content and COD elimination rate of the pilot plant related to the mixed feed without glucose dosing

in the PA, as with COD, is also inert, i.e. biologically non-degradable or hardly degradable. This part is rather due to the refractory organic nitrogen compounds in pharmaceutical production (Fig. 4.23).

For the parameters NH4-N, BSB5 and SS the plant delivers fully satisfactory results, and due to the somewhat low concentrations in the PA feed, the P elimination with an average effluent value of 0.54 mg/l has been much improved compared to project phase T1. The average elimination rates as well as inlet and outlet concentrations of all parameters are shown in the following table. The COD elimination rate here refers to the mixed feed with C-dosage. For the calculation of the data, representative time periods were selected in which the PA content does not exceed 28% and is not influenced by previous test series (e.g. between day 26 and 69).

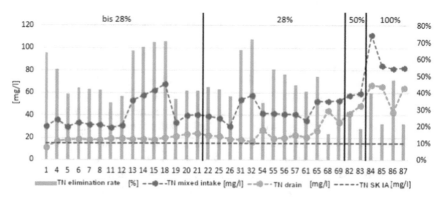

Fig. 4.23 Inflow and effluent concentrations of TN as a function of PA content during the operating phase MBR

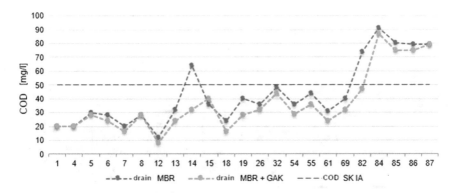

Fig. 4.24 Change in COD concentrations due to downstream GAK filter

MBR + GAK

The Granulated Activated Carbon Filter (GAK) was also put into permanent operation during project phase T2. In order to be able to assess the effectiveness of the GAK in direct comparison to the plant without GAK, Fig. 4.24 shows the COD discharge values for the same measuring days for the above described operating phase "MBR" and for "MBR + GAK". On a total of 23 of the 34 days of the "MBR" operating phase, the process could be analysed according to GAK. This clearly shows that the GAK filter further increases the cleaning performance. Only on one day the COD discharge of the filter is above that of the test plant. This value can be justified by measurement inaccuracy, as the values only differ from one another by 4 mg/L. It can also be seen that even the GAK filter cannot significantly reduce the inert COD content any further, especially if the PA content is increased to 100%. The average COD elimination rate between the MBR filtrate and the GAK drain for the measuring days is 18.26% (without measuring day 15).

The average COD concentration of the mixed feed, the average COD effluent concentration of the MBR (filtrate) and MBR + GAK and the average COD elimination rate in relation to the mixed feed with glucose are shown in Table 4.18 below (COD in relation to the mixed feed with C source). In order to achieve uniformity of presentation, the operating phase "Increase in proportion of the PA" is also not included in the statistics here. It can be seen that the elimination rate can be increased by approx. 2% by using the GAK. In relation to the mixed feed without glucose, it is even almost 5% (Table 4.17).

Table 4.18 shows the elimination rates as well as inlet and outlet concentrations for the measured parameters. For TP and TN no significant improvement could be found by using GAK and due to the workload of the laboratory not many measured values could be collected, which limits the data reliability.

Preozoning + MBR + GAK

During pre-ozoning, the PA partial water stream was ozonated before it merged with the municipal wastewater partial stream. The exact dosing point is shown in Fig. 4.9.

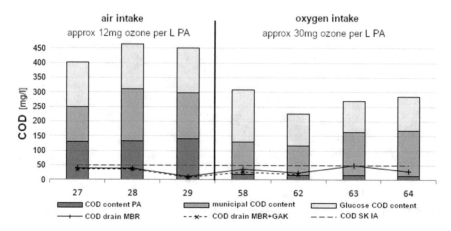

Fig. 4.25 Inflow and effluent concentrations of the operating phase preozoning + MBR + GAK

The original aim of the pre-ozonation of the PA partial flow is that a favourable substrate supply (improved BOD:COD or BOD:N ratio in the inflow) for the microorganisms in the aeration tank can be achieved by splitting up the hardly degradable molecules and compounds in the PA. The ozone was added to the PA partial water stream via a diffuser while it was pumped from the flocculation tank through a pipe in the direction of the confluence of the two partial water streams. In the first series of tests with air supply, the dosing quantity of ozone was fixed at approx. 12 mg per litre of feed PA on the basis of preliminary tests carried out in the laboratory, and in the second series of tests with oxygen supply it was increased to approx. 30 mg/l. The COD inlet and outlet values of the plant are shown in Fig. 4.25. In addition, samples were taken on some days not only from the filtrate but also from the GAK filter, and these values are also shown.

From the 7-day trials it can be seen that the limit value of 50 mg/L can always be complied with. It is noticeable that the effluent values do not look particularly better in comparison to operating phases without ozonation and that the cleaning performance cannot be increased with higher ozone dosage. It can be assumed that also the municipal wastewater partial flow has inevitably come into contact with ozone. This contact takes place in the inlet pipe to the coarse material cell and also in the coarse material cell. The total contact time of the ozone with the wastewater flow is difficult to assess and, according to all known information, is about 15 to 20 s in the pipe to the coarse material cell and then in the coarse material cell at seconds to a few minutes, while the actual reaction time between ozone and wastewater constituents should be less. Sampling after the theoretical contact time for testing the immediate effect of ozone is unfortunately not possible due to the practical structure of the system.

The COD inlet and outlet values in relation to the mixed inlet with C-source are listed in the following Table 4.19. As a result of the fluctuations in the feed, somewhat larger deviations in the discharge can also be found. For the same 5 measuring days an increase of the elimination rate of about 2% was achieved by the GAK filter.

MBR + Postozoning + GAK

During post-ozoning, ozone was added to the filtrate (MBR effluent) in the filtrate tank. The exact point of dosing is shown in Fig. 4.9. The aim was to decompose the inert COD or the COD not yet decomposed in the biological treatment stage into filterable "fragments" by ozonation. There may also be pharmaceutical and other trace substances in the plant effluent which can be destroyed by oxidation. The resulting "fragments" can then be filtered out through the GAK filter if necessary.

The contact time of the water with the ozone was on average about 5 min. The excess ozone was removed from the container via a vent pipe. Based on practical experience in Germany and China, different dosage quantities were tested, which varied between 3 and approx. 20 mg/l filtrate depending on the source of supply. Figure 4.26 shows the COD inflow and outflow values and the corresponding ozone quantity (n = 8, outflow values from different sampling points are also shown on a large scale).

It can be seen that the limit value of 50 mg/l can always be complied with. However, the COD discharge value cannot be reduced constantly by using post-ozoning alone. As with pre-ozoning, a significant increase in the ozone quantity does not result in a higher purification performance in terms of COD. In this case, the downstream GAK filter leads to the separation of a fairly large proportion of the remaining COD at an increased dosing quantity. On average, the use of the GAK filter reduces the incoming COD concentration of the already ozonated filtrate of the membrane stage by about 20%. The average COD elimination rate between the plant outlet not yet treated with ozone and the outlet after the GAK filter is even almost 30%. This is due to the fact that the ozone makes the inert COD portion filterable by splitting.

The average COD elimination levels, as well as inflows and outflows, are summarized in Table 4.20. It is clear that the effluent concentration can be further reduced with each stage - especially by combining it with the GAK filter. However, even here the number of measured data is not sufficient to draw a meaningful conclusion due to the limited test time.

MBR + PAK (+GAK)

In order to make more powerful statements about the effect of dosing PAHs on the cleaning performance of the test plant, an additional series of tests was carried out in which only PAHs were added to the plant as dosing agents. The results of the tests are presented in this chapter.

The PAH was dissolved in water (100 g/l) in a canister and added to the revitalisation by means of a peristaltic pump. The exact dosing point is shown in Fig. 4.9 or in the picture in the appendix. To prevent the PAHs from settling in the PAH solution, the solution was constantly mixed in the canister with the aid of aeration bars.

According to [15], to reduce the COD concentration from 50 mg COD/l to 20 mg COD/l, 20 mg PAH/l must be dosed. The average COD withdrawal in their investigations was 0.8–1.2 g COD/g PAH. Different dosages of PAH were investigated in

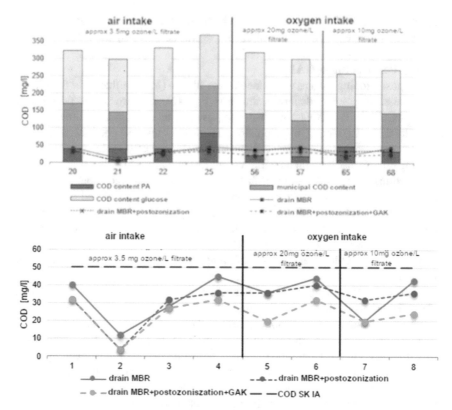

Fig. 4.26 Inflow and effluent concentrations of COD of the operating phase MBR + post-ozoning + GAK

the operating phase, varying between 10 and 50 mg/l. Figure 4.27 shows the COD inflow and outflow values and the dosing quantities of PAHs per litre of mixed feed (n = 7). The corresponding operating conditions are listed in the appendix.

Due to the time limit, only 2 to a maximum of 3 measuring days could be carried out for each dosing variant. During this operating phase, the plant was fed with PA batches that had already been subjected to the membrane stage and were presumably difficult to degrade. These batches are characterised by constantly low COD concentrations and poor BOD5:N ratios. It is noticeable that the significant increase in the PAH dosage did not lead to any improvement and that the limit value for COD of 50 mg/l in the filtrate could not be met on measurement day 80. However, the downstream GAK filter means that the limit value is undercut.

It can be assumed that trace substances that have accumulated on the large inner surface of the PAHs have been removed together with the PAHs with the ODP. However, the dry matter content of the sludge in the aeration tanks and the filter chamber has increased significantly (by approx. 5 g/l each) due to the use of

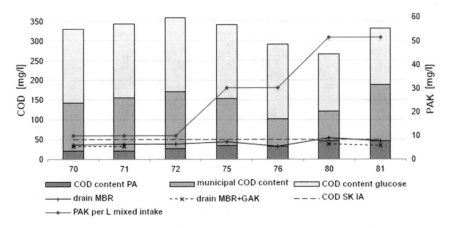

Fig. 4.27 COD inflow and outflow values of the operating phase MBR + PAH (+ GAK)

PAHs. The optimum values for the sludge parameters could therefore no longer be maintained for the short-term measures.

The following table lists all measured parameters in the mixed feed (with glucose) and the respective processes with the elimination rates. The lower purification performance for TP and TN compared to the FM operating phase indicates that PAH alone as an additional stage in combination with MBR cannot increase the plant performance in terms of nutrient elimination (Table 4.21).

In addition to the dosing of PAH, the pre-ozonation of the PA partial water stream was tested on one day and the post-ozonation of the MBR filtrate on one day. Since only one day was tested for each type of ozonation, no representative conclusions can be drawn from the results. It can be stated that no major COD cleaning performance was achieved by using both ozonation variants.

4.1.2.4.3 Comparison and Evaluation of Results

The results clearly show that the use of the GAK filter causes the most significant increase in COD cleaning performance. If we take COD measured values before and after the GAK filter every day, we obtain an average additional increase in the degree of elimination of approx. 20% (n = 53). This means that the COD discharge values can be reduced by about one fifth. This also includes all measurement days on which other processes were tested (ozonation, FM, FHM, PAH).

The results of the sole use of the test facility (MBR stage) and the combination of the MBR stage and the GAK filter (n = 23) are shown in the following figure in the form of box plots, so that strongly deviating values have less influence on the representation. This illustration also clarifies the statement that the cleaning performance can be increased by using GAK filters, as the median (the line dividing the box) of the "MBR + GAK" combination is 8 mg/l below the median of the MBR stage alone (Fig. 4.28).

To go into the COD elimination further, it can be stated that in addition to the use of the GAK filter, the use of other cleaning stages also produced good results.

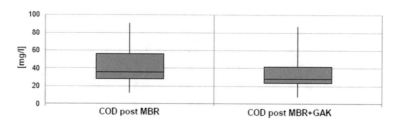

Fig. 4.28 Boxplots for effluent concentrations of COD before and after GAK filtration

Unfortunately, it was not possible to collect a lot of data for some measurement series (ozonation, FM, PAH), so that the statements made on this should be confirmed in further tests. Due to the great variation in PA quality between different PA batches (see Fig. 4.18 and Fig. 4.19), a direct comparison of the absolute effluent concentrations of all test series is not of great importance. Nevertheless, the following diagram shows the average COD elimination levels of each test series. In order to be able to assess the significance of the results, the number of measured values for the respective test series is also given. The combination "FM + MBR + PAH + GAK" shows (for a small test series) the best elimination levels, which is probably due to the extremely high inlet concentrations. In case of very high COD inlet values, slight increases in the degrees of elimination cause big differences in the outlet values. This should be taken into account when planning new wastewater treatment plants in the region. It is noticeable that the treatment performance of the "MBR + PAH + GAK" test series is decreasing compared to others. Even with GAK, the average rate does not reach that of the "MBR" test series. The reason for this is, as already clarified, the commissioning of the membrane stage in NEPG, whose effluents from that time on contained hardly degradable COD and only had a negative dilution effect on the entire feed (Fig. 4.29).

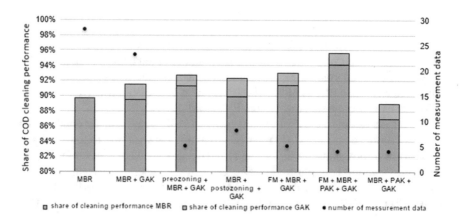

Fig. 4.29 Mean COD elimination rates for the different pilot phases.

Table 4.16 Inflow and effluent concentrations and elimination rates of the operating phase MBR in T2

Parameters		Mixed feed [mg/l]		Outflow [mg/l]		Elimination rate [%]	
Description	n	Ø	s	Ø	s	Ø	s
COD	28	313,4	43,2	32,3	12,0	89,7%	4,6%
NH₄-N	26	23,7	5,6	0,10	0,04	99,6%	16,1%
TN	22	41,2	11,0	19,4	6,0	49,9%	14,8%
TP	20	2,9	0,4	0,54	0,2	81,3%	8,0%
SS	22	87,7	19,4	0,07	0,25	99,9%	0,2%

The results of the test series with ozonation had not shown in the relatively short test period that ozone can act more effectively than activated carbon for the treatment of the wastewater streams in Shenyang. For economic and safety reasons, activated carbon (in both forms) could be of more importance for large-scale industrial plants. In general, the pilot plant was able to eliminate at least 87% of the COD loads in the influent with a PA content of up to 30% under all possible process combinations. The limit value of 50 mg/l of the SK IA could be maintained on almost all measuring days even without GAK or other processes (but with C-dosage).

As can be seen from Table 4.16, the use of the pilot plant without any further processes has made it possible to achieve cleaning performances for COD, BOD5, NH4-N, TP and SS that meet the highest effluent standard IA from Table 4.2. For BOD, NH4-N and SS, the standards are far below those shown in Table 4.2. The limit value of TP (0.50 mg/l) is only slightly exceeded (0.52 mg/l). For the TP effluent values, it should be noted that the use of FM has resulted in an increase in cleaning performance. The addition of FM did not result in significantly lower TP values. Here the cleaning performance is increased by just 1.45%. In general, it can be said that the use of FM is a possibility to further reduce both TP and COD effluent values (especially for inert COD contents). This can be seen in the test results presented as well as in the results from the preliminary tests for FM selection, where the use of the PAC has already achieved a COD elimination level of 20% and a TP elimination level >80%.

Only the limit value for TN cannot always be complied with and is exceeded by almost 30%. The reasons for this are, as explained several times, the oxygen carry-over from the filter chamber into the denitrification tank and the poor substrate supply in the inlet. In order to support the degradation of TN, glucose was added as a

Table 4.17 Mixed feed and effluent concentrations and elimination rates of COD for the operating phases MBR and MBR + GAK

Procedure		Mixed feed [mg/l]		Outflow [mg/l]		Elimination rate [%]	
Description	n	Ø	s	Ø	s	Ø	s
MBR	18	311,1	49,2	32,7	11,8	89,5%	3%
MBR + GAK	18	311,1	49,2	26,7	8,7	91,5%	2%

Table 4.18 Mixed feed and effluent concentrations and elmination rates for different parameters of the operating phase MBR + GAK

Parameters		Mixed feed [mg/l]		Outflow [mg/l]		Elimination rate [%]	
Description	n	Ø	s	Ø	s	Ø	s
COD	18	311,1	49,2	26,7	8,7	91,5%	2%
TN	3	42,83	8,0	20,20	0,9	50,1%	10,4%
TP	3	2,52	0,5	0,37	0,2	86,3%	7,3%
SS	18	72,63	39,2	0,00	0	100%	0

Table 4.19 Mixed feed and effluent concentrations and elimination rates of COD in the operation phase pre-ozonation + MBR + GAK

Procedure		Mixed feed [mg/l]		Outflow [mg/l]		Elimination rate [%]	
Description	n	Ø	s	Ø	s	Ø	s
Preozoning + MBR	7	343,8	87,1	33,0	11,5	89,7%	4,3%
Preozoning + MBR + GAK	5	370,1	90,5	25,6	10,6	92,7%	2,8%

Table 4.20 Mixed feed and effluent concentrations and elimination rates of COD for the operation phases MBR, MBR + post-ozoning and MBR + post-ozoning + GAK

Procedure		Mixed feed [mg/l]		Outflow [mg/l]		Elimination rate [%]	
Description	n	Ø	s	Ø	s	Ø	s
MBR	8			33,5	11,5	89,1%	3,7%
MBR + Postozoning	8	307,5	32,9	31,0	10,5	89,8%	3,6%
MBR + Postozoning + GAK	8			23,9	8,9	92,3	2,7%

Table 4.21 Mixed feed and effluent concentrations and elimination rates of COD, TN and TP for the operation phase MBR + PAH (+ GAK)

Procedure			Mixed feed [mg/l]		Outflow [mg/l]		Elimination rate [%]	
Description		n	Ø	s	Ø	s	Ø	s
COD	MBR + PAK	7	323,1	30,6	41,6	6,8	86,9%	3,2%
	MBR + PAK + GAK	4	317,8	30,1	34,5	7,0	88,9%	2,1%
TN	MBR + PAK	7	48,1	3,2	24,1	3,3	49,8%	6,7%
TP	MBR + PAK	5	3,4	0,1	0,4	0,1	88,3%	3,5%

C source to varying degrees (setting C:N ratio), which had a great effect to a certain extent. However, the TN feed values of the PA were subject to strong fluctuations, so that the adjustment of the dosing quantity could often only be carried out too late - as soon as the measured values were announced by the laboratory. Figure 4.30

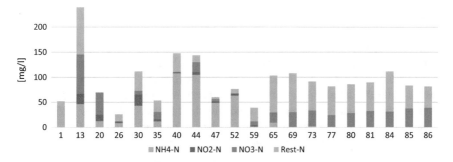

Fig. 4.30 Composition of nitrogen in the PA over the entire project period

shows the measured nitrogen fractions in PA of all batches. Significant changes in composition are only seen from the start of operation of the membrane stage of NEPG with a dominance of nitrate and residual N, whereas before that there are strong fluctuations. The large proportion of organic residual N also confirms the poor degradability of TN in PA.

In order to investigate the effect of C-dosage or denitrification to a large extent, nitrite and nitrate concentrations were also measured in the effluent (filtrate) on certain days. Figure 4.31 shows that the nitrate fraction had been decomposed properly under C-dosage independent of test series and PA content, with nitrite and NH4-N hardly being found. The dosing ratio (BOD5: N 4–5) tested was sufficient to bring the nitrate concentration below 10 mg/l for most of the measuring days, even in the presence of oxygen carry-over. Not degradable are the residual N-fractions existing in the mixed feed, which are organic or inert and cannot be biologically ammoniated. For the elimination of this fraction further approaches are necessary.

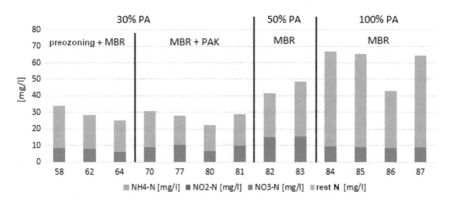

Fig. 4.31 Composition of nitrogen in the effluent (filtrate) over the entire project period.

4.1.2.4.4 Recommendation for Action on Pharmaceutical Wastewater Treatment in Shenyang

The main objective of subproject T2 is to develop a recommendation for the treatment of pharmaceutical (and municipal) wastewater for the Shenyang region.

From a purely process-technical point of view, it can be interpreted from the test results that for a locally planned mixed treatment of different types of wastewater, in addition to the use of the membrane activated sludge process, above all the downstream connection of activated carbon filters is to be used. Not only COD effluent concentrations can be reduced in this way, but also trace material loads. The recommendation to use GAK filters can be supported by a large amount of data, in contrast to other test series.

The results of the test series on ozonation also provide high COD elimination rates. However, ozone is not easy to handle and the energy costs for ozone production in ozone generators are also high. The corrosive effect has already been shown after a few days in the test plant on the ozone system, when leaks occurred in connections and hoses. This leads to extensive maintenance work and associated costs. In addition, unwanted oxidation products can be produced during ozonation. In this case, precise investigations are necessary for concrete plant planning. It is estimated that the micro pollutants introduced in Germany a few years ago will not be regarded as standard parameters in China in the near future, so that the use of ozonation is not recommended as a first priority compared to GAK filters.

The dosing of external C-source can be used to adjust the C:N ratio at high TN feed values. Since the majority of the COD in pharmaceutical wastewater is most likely inert, dosing of readily degradable carbon in the form of glucose or similar will still be useful to support the degradation of TN and TP. However, the cost and effort factor must also be taken into account here. For example, intermittent dosing is recommended, which should be based on the results of preliminary tests and online monitoring of the COD and nitrate concentrations in the mixed feed or in the activated sludge.

The highest COD elimination rates were achieved by a combination of pre-precipitation of PA, treatment in the MBR pilot plant, dosing of PAH into the revitalisation system and downstream connection of a GAK filter. Due to the cost factor and the complexity of mixing and dosing, only one of the two dosing agents should be selected in large-scale technology (and if necessary only for emergencies). It should be noted that PAH can be dosed in powder form. As in subproject T2, preliminary tests are useful for selecting a suitable agent. For the pharmaceutical wastewater of NEPG in Shenyang, the use of flocculants can be dispensed with, since no or only low elimination rates were achieved (COD: no increase; TP: 1.45%).

According to the collected and analysed results, the combination of a possible pre-precipitation for heavily loaded wastewater substreams (especially with high COD and TP inflow values with online monitoring as a prerequisite), a possible simultaneous precipitation as well as glucose dosage into the activation, the membrane activated sludge process as main component and a downstream activated carbon filter is proposed as a recommendation for the co-treatment of pharmaceutical wastewater in municipal wastewater treatment plants in Shenyang.

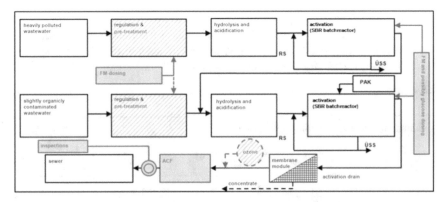

Fig. 4.32 Schematic diagram of the NEPG wastewater treatment plant

As a concrete example of the technical upgrade of an industrial wastewater treatment plant, the NEPG wastewater treatment plant, after an external membrane module has already been installed, can be further modified by treatment with FM and the use of activated carbon filters in case of emergency. After thorough preliminary investigations, ozonation or PAH dosage can also be carried out in an emergency. The technical and economic feasibility must be checked in the course of this and further investigations and tests must be carried out with the actual wastewater. Fig. 4.32 shows a schematic diagram of the NEPG's operational wastewater treatment plant with the technical equipment (fields marked green).

Obviously, there is a great variation in the material loads of the delivered wastewater batches of NEPG, which can be traced back to both the functioning of the operational pre-treatment stages and the production of NEPG. Throughout the entire project period, different production processes were running in the pharmaceutical plant, the cycle of which is unknown for reasons of confidentiality. The fluctuations are not only to be found in NEPG, but also in experience with other plants that have no pre-treatment or only limited pre-treatment. Therefore, beyond the procedural recommendations for action, stricter controls of the discharging industrial plants as well as the technical upgrading of the operational wastewater treatment plants should be addressed in order to ensure that the set discharge standards into the sewerage system (see Table 4.5) can be met.

However, as can be seen from the results (also from T1), a large amount of inert COD is discharged into the public sewerage system by NEPG or other companies. The inert COD cannot be biodegraded at all or only to a small extent. Furthermore, the nutrient ratio in industrial wastewater is not suitable for the degradation of nitrogen, a typical example being the extremely poor C:N ratio in pharmaceutical wastewater after the use of NEPG's own membrane stage. The cleaning performance of the public sewage treatment plant is impaired by the dilution effect of industrial partial flows, even though the discharge standard of the indirect dischargers is maintained.

In this context, it is particularly recommended to adapt the Chinese discharge standard of plants where not only limit concentrations for certain parameters have to be observed. The discharging companies should ensure a nutrient ratio in their already pre-treated wastewater by adapting the process technology so that municipal wastewater treatment plants have better basic conditions for nitrogen and phosphorus degradation. Further nutrient elimination in industrial wastewater.

4.1.4 Annex

sampling protocol pilot plant Shenyang

sampler: Maximilian Roß
weather: sunny
date: 24.11.17
flow MBR (l/h): 320

sampling location	municipal intake	drain without ozone	nitrification	filter chamber	drain after ACF	pharma intake
sampling type	spot check	mixed check	spot check	spot check	spot check	spot check
time	08:55	9:00; 13:00; 15:00	09:10	09:05	15:10	15:30
on-side parameters						
temperature (°C)	14,9	17,5; 18,0; 16,3				26,5
PH value	7,65	7,77				7,63
conductivity	1581	1397				1797
oxigen content						0,16
visual coloring	light grey	clear	dark brown	dark brown	clear	grey
cloudiness	heavy	none	heavy	heavy	none	heavy
smell	putry, jauchy, sweet	lightly earthy	earthy	earthy	lightly earthy	purty (rotten eggs)
Pretreatment		cooled			cooled	cooled
examined parameters						
COD(mg/L)	198	20			20	182
gel.COD (mg/l)						
BOD5 (mg/L)						
NH4-N (mg/L)	22,75	0,12				43
Ntot (mg/L)	28,4	10,8				52
Ptot (mg/L)	2,77	0,08				2,93
TR (%)						0,1027
oTr (%)						0,0355
SS	128	1				84
TS (mg/L)			9624	12344		
SV30 (ml/L)			1000	1100		
comment						after 15:30: 2% Level from flocculation tank every 120 min

Annex 4-1: Sampling protocol

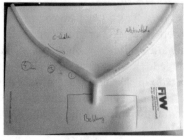

Annex 4-2: Dosing tank for glucose and powdered activated carbon (top left), dosing hose (bottom left) and view into the experimental container

Annex 4-3: ozonegenerator „C-Lasky DTI" by „AirTree" (FiW)

Annex 4-4: flocculation tank (inside) and agitator (FiW)

Annex 4-5: adjustable operating parameters inculding scale

Pre-cleaning	Finescreen	Feeding pump 1	Stop	Auto	By hand	
		drive Finescreen	Stop	Auto	By hand	
		Dry running	0 – 100 %			
		Rake off	0 – 100 %			
		Rake on	0 – 100 %			
		High alarm	0 – 100 %			
		Delay time	0 – 999 s			
		Pace on	0 – 999 s			
		Pace off	0 – 999 s			
		Max. runtime	0 – 999 h			
	pharmaceutical waste-water	Pharma transfering pump	Stop	Auto	By hand	
		Dosing pump FM	Stop	Auto	By hand	
		runtime FM-pump	0 - ∞ s			
		Dosing pump FHM	Stop	Auto	By hand	

		Runtime FHM pump	0 - ∞ s			
		Agitator	Stop		Auto	By hand
		Mixing time	0 – 999 h			
		Feeding pump 2	Stop		Auto	By hand
		Dry running	0 – 100 %			
		Pump off	0 – 100 %			
		Start of flucculation	0 – 100 %			
		High alarm	0 – 100 %			
		NOR feeding pump 2	0 – 100 %			
		Total time	0 – 999 min			
Activation	Denitrification	Compressor denitrification	Stop		Auto	By hand
		Pace on	0 – 999 s			
		Pace off	0 – 999 s			
	Nitrification	Compressor nitrification	Stop		Auto	By hand
		Pace on	0 – 999 s			
		Pace off	0 – 999 s			
		Min air flow	0 – 99 Nm^3/h			
		Max air flow	0 – 99 Nm^3/h			
	Activation tank	Dry running	0 – 100 %			
		Filtration off	0 – 100 %			
		Filtration in	0 – 100 %			
		Feeding pump 1 stop	0 – 100 %			
		Feeding pump 2 stop	0 – 100 %			
		High alaerm	0 – 100 %			
	Ball valves	Recirculationpump	Stop		Auto	By hand
		EKH M10	Stop		Auto	By hand
		EKH M11	Stop		Auto	By hand
Filtration	Compressor filtration	Compressor 1 filtration tank	Stop		Auto	By hand
		Compressor 2 filtration tank	Stop		Auto	By hand
		Pace on	0 – 999 s			
		Pace off	0 – 999 s			
		Min air flow	0 – 99 Nm^3/h			
		Max air flow	0 – 99 Nm^3/h			
	Filtration	Filtrate pump	Stop		Auto	By hand
		Dry running	0 – 100 %			
		Filtrate amount	0 – 3000 L/h			
		Filtrate amount min.	0 – 3000 L/h			
		p_min	0 – (-350) mbar			
		p_minmin	0 – (-350) mbar			
		Pace on	0 – 999 s			

		Pace off	0 – 999 s		
	Filtrattank	Manual set point	0 – 100%		
		Filtrate transfer pump	Stop	Auto	By hand
		Pump off	0 – 100 %		
		Pump on	0 – 100 %		
		High alarm	0 – 100 %		
discharge and regulation	Sludge discharge	Setpoint sludge discharge	0 – 100 m³		
		Actual value sludge discharge	-		
		Parts sludge	0 – 5 %		
	Coarse	Setpoint coarse-discharge	0 – 200 m³		
		Actual value coarse-discharge	-		
		Chemical cleaning	on		off

Annex 4-6: Used sensors and probes and their tasks

Sensor/Probe	Installation point	Purpose
Level sensor	Coarse material cell, nitrification / denitrification basin, filter chamber	Operational flow
Temperature sensor	Nitrification / denitrification basins	Biodegradation monitoring
Oxygen content probe	Nitrification basin	Assessment of the ventilation setting
Redox potential probe	Denitrification basin	Assessment of denitrification performance
Pressure sensor	Between the diaphragm and the diaphragm pump	Monitoring the condition of the membrane

Annex 4-7: Volume of the plant components of the test plant

Investment component	Description	Average filling level	Volume	Efficient Volume
Filter chamber	V_F	91%	2 m³ —0,8 m³	1,2 m³
Aeration tank (denitrification: nitrification)	V_{BB} ($V_{Deni.}$:$V_{Nitri.}$)	67% - 75%	1,4 m³ (1:1)	1 m³
Coarse matter cell	V_G	100%	approx. 0,8 m³	0,8 m³
Activated carbon filter	V_{AKF}	-	31,3 L each	1,3 L each
Filtrate tank	V_P	50%	approx. 60 L	approx. 60 L
Pipes (length: 30 m)	V_R	-	approx. 60 L	approx. 60 L

Determination SV30

First the stand cylinder is filled with 800 ml MBR wastewater (process water). Then the sample to be tested is removed from the filter chamber or nitrification basin. Before the sample can be taken from the nitrification, the aeration of the nitrification must be switched on for 2 min to ensure that the nitrification sludge is mixed. Afterwards, the cylinder is filled up with 200 ml sludge, so that the volume of the filtrate-sludge-mixture is 1 L. By closing the cylinder and the subsequent mixing by shaking, a complete mixing in the cylinder is achieved. Now a 30-min settling time is waited for. The sludge should now have settled so that a clear and a sludgey phase can be clearly seen. The read value of the sludge phase (in ml) must now be multiplied by 5 to obtain the SV30.

Annex 4-8: Allocation of the calendar dates to the diagram numbers for project phase T1

Nr.	date	Nr.	date	Nr.	date	Nr.	date
1	04.08.2017	32	04.09.2017	63	05.10.2017	94	05.11.2017
2	05.08.2017	33	05.09.2017	64	06.10.2017	95	06.11.2017
3	06.08.2017	34	06.09.2017	65	07.10.2017	96	07.11.2017
4	07.08.2017	35	07.09.2017	66	08.10.2017	97	08.11.2017
5	08.08.2017	36	08.09.2017	67	09.10.2017	98	09.11.2017
6	09.08.2017	37	09.09.2017	68	10.10.2017	99	10.11.2017
7	10.08.2017	38	10.09.2017	69	11.10.2017	100	11.11.2017
8	11.08.2017	39	11.09.2017	70	12.10.2017	101	12.11.2017
9	12.08.2017	40	12.09.2017	71	13.10.2017	102	13.11.2017
10	13.08.2017	41	13.09.2017	72	14.10.2017	103	14.11.2017
11	14.08.2017	42	14.09.2017	73	15.10.2017	104	15.11.2017
12	15.08.2017	43	15.09.2017	74	16.10.2017	105	16.11.2017
13	16.08.2017	44	16.09.2017	75	17.10.2017	106	17.11.2017
14	17.08.2017	45	17.09.2017	76	18.10.2017	107	18.11.2017
15	18.08.2017	46	18.09.2017	77	19.10.2017	108	19.11.2017
16	19.08.2017	47	19.09.2017	78	20.10.2017	109	20.11.2017
17	20.08.2017	48	20.09.2017	79	21.10.2017	110	21.11.2017
18	21.08.2017	49	21.09.2017	80	22.10.2017	111	22.11.2017
19	22.08.2017	50	22.09.2017	81	23.10.2017	112	23.11.2017
20	23.08.2017	51	23.09.2017	82	24.10.2017	113	09.01.2018
21	24.08.2017	52	24.09.2017	83	25.10.2017	114	10.01.2018

22	25.08.2017	53	25.09.2017	84	26.10.2017	115	11.01.2018
23	26.08.2017	54	26.09.2017	85	27.10.2017	116	12.01.2018
24	27.08.2017	55	27.09.2017	86	28.10.2017	117	13.01.2018
25	28.08.2017	56	28.09.2017	87	29.10.2017	118	14.01.2018
26	29.08.2017	57	29.09.2017	88	30.10.2017	119	19.02.2018
27	30.08.2017	58	30.09.2017	89	31.10.2017	120	20.02.2018
28	31.08.2017	59	01.10.2017	90	01.11.2017	121	21.02.2018
29	01.09.2017	60	02.10.2017	91	02.11.2017	122	22.02.2018
30	02.09.2017	61	03.10.2017	92	03.11.2017	123	23.02.2018
31	03.09.2017	62	04.10.2017	93	04.11.2017		

Annex 4-9: Allocation of the calendar dates to the diagram numbers for project phase T2

1	24.11.17	23	16.12.17	45	07.01.18	67	04.02.18
2	25.11.17	24	17.12.17	46	08.01.18	68	05.02.18
3	26.11.17	25	18.12.17	47	15.01.18	69	06.02.18
4	27.11.17	26	19.12.17	48	16.01.18	70	07.02.18
5	28.11.17	27	20.12.17	49	17.01.18	71	08.02.18
6	29.11.17	28	21.12.17	50	18.01.18	72	09.02.18
7	30.11.17	29	22.12.17	51	19.01.18	73	10.02.18
8	01.12.17	30	23.12.17	52	20.01.18	74	11.02.18
9	02.12.17	31	24.12.17	53	21.01.18	75	12.02.18
10	03.12.17	32	25.12.17	54	22.01.18	76	13.02.18
11	04.12.17	33	26.12.17	55	23.01.18	77	14.02.18
12	05.12.17	34	27.12.17	56	24.01.18	78	24.02.18
13	06.12.17	35	28.12.17	57	25.01.18	79	25.02.18
14	07.12.17	36	29.12.17	58	26.01.18	80	26.02.18
15	08.12.17	37	30.12.17	59	27.01.18	81	27.02.18
16	09.12.17	38	31.12.17	60	28.01.18	82	28.02.18
17	10.12.17	38	01.01.18	61	29.01.18	83	01.03.18
18	11.12.17	40	02.01.18	62	30.01.18	84	02.03.18
19	12.12.17	41	03.01.18	63	31.01.18	85	05.03.18
20	13.12.17	42	04.01.18	64	01.02.18	86	06.03.18
21	14.12.17	43	05.01.18	65	02.02.18	87	07.03.18
22	15.12.17	44	06.01.18	66	03.02.18	88	08.03.18

Annex 4-10: Data of the siClaro® FM 622 and the MBR (MMS, o. J.)

	siClaro® FM 622	MBR
Filtration area	25 m^2	50 m^2
Dimensions (L x W x H)	608 x 362 x 1360 [mm]	2 Modules of this size
max. flowrate	11 l/(m^2*h)	22 l/(m^2*h)
max. flow	14,700 m^3/d	29,400 m^3/d
max. Membrane pressure	-300 mbar	-300 mbar
Purge air requirement	16,500 Nm^3/h	33,000 Nm^3/h

Annex 4-11: Operating conditions and operating parameters for all T2-related test phases

Operating phase MBR

Operating parameters	n	average	s	Mud age [d]	34	20,56	2,58
Q_N [l/h]	34	292,61	57,34	TS_{BB} [g/l]	34	11,98	1,65
Q_B [l/h]	34	365,76	71,67	$SV30_{BB}$ [ml/l]	34	1026	125
MF_N [l/(m^2*h)]	34	6,06	1,10	$B_{R,COD}$ [kg/(m^3*d)]	34	1,10	0,27
MF_B [l/(m^2*h)]	34	7,57	1,38	$B_{R,BOD5}$ [kg/(m^3*d)]	34	0,59	0,21
P_N [l/(m^2*h*bar)]	34	69,86	23,04	$B_{TS,COD}$ [kg/(kg$_{TS}$*d)]	34	0,088	0,044
P_B [l/(m^2*h*bar)]	34	87,32	28,80	$B_{TS,BOD5}$ [kg/(kg$_{TS}$*d)]	34	0,048	0,025
pH	34	7,52	1,19	COD:N	34	3,83	1,69
O_2 [mg/l]	34	2,18	0,55	(COD:N)$_G$	34	7,65	2,12
T_{BB} [°C]	30	18,48	0,90	BOD_5:N	34	1,35	0,96
RV	34	6,18	1,86	(BOD_5:N)$_G$	34	4,33	1,43

MBR + GAK

Operating parameters	n	average	s	Mud age [d]	23	22,14	3,11
Q_N [l/h]	23	296,81	62,66	TS_{BB} [g/l]	23	11,97	1,79
Q_B [l/h]	23	371,01	78,33	$SV30_{BB}$ [ml/l]	23	1026	125
MF_N [l/(m^2*h)]	23	6,15	1,12	$B_{R,COD}$ [kg/(m^3*d)]	23	1,15	0,27
MF_B [l/(m^2*h)]	23	7,68	1,40	$B_{R,BOD5}$ [kg/(m^3*d)]	23	0,65	0,22
P_N [l/(m^2*h*bar)]	23	72,28	25,96	$B_{TS,COD}$ [kg/(kg$_{TS}$*d)]	23	0,086	0,030
P_B [l/(m^2*h*bar)]	23	90,35	32,45	$B_{TS,BOD5}$ [kg/(kg$_{TS}$*d)]	23	0,049	0,019
pH	23	7,44	1,43	COD:N	23	3,75	1,90
O_2 [mg/l]	23	2,21	0,55	(COD:N)$_G$	23	7,75	2,29
T_{BB} [°C]	19	18,54	0,76	BOD_5:N	23	1,39	0,83
RV	23	7,14	1,65	(BOD_5:N)$_G$	23	4,49	1,32

Preozoning + MBR + GAK

Operating parameters	n	average	s	Mud age [d]	7	26,84	3,01
Q_N [l/h]	7	239,74	45,89	TS_{BB} [g/l]	7	12,65	0,91
Q_B [l/h]	7	299,68	57,37	$SV30_{BB}$ [ml/l]	7	1021	99
MF_N [l/(m^2*h)]	7	4,86	0,99	$B_{R,COD}$ [kg/(m^3*d)]	7	0,97	0,41
MF_B [l/(m^2*h)]	7	6,07	1,24	$B_{R,BOD5}$ [kg/(m^3*d)]	7	0,45	0,25

P_N [l/(m²*h*bar)]	7	64,01	18,78	$B_{TS,COD}$ [kg/(kg$_{TS}$*d)]	7	0,074	0,037
P_B [l/(m²*h*bar)]	7	80,01	23,47	$B_{TS,BOD5}$ [kg/(kg$_{TS}$*d)]	7	0,034	0,022
pH	7	7,70	0,22	COD:N	7	6,36	2,88
O_2 [mg/l]	7	1,61	0,54	(COD:N)$_G$	7	10,48	3,67
T_{BB} [°C]	7	19,01	1,24	BOD_5:N	7	2,21	1,78
RV	7	6,46	2,69	(BOD_5:N)$_G$	7	5,40	2,35

MBR + postozoning + GAK

Operating parameters	n	average	s	Mud age [d]	8	23,39	3,53
Q_N [l/h]	8	273,34	42,76	TS_{BB} [g/l]	8	12,07	1,42
Q_B [l/h]	8	341,67	53,45	$SV30_{BB}$ [ml/l]	8	1044	136
MF_N [l/(m²*h)]	8	5,56	0,92	$B_{R,COD}$ [kg/(m³*d)]	8	0,93	0,19
MF_B [l/(m²*h)]	8	6,95	1,15	$B_{R,BOD5}$ [kg/(m³*d)]	8	0,45	0,15
P_N [l/(m²*h*bar)]	8	67,54	13,95	$B_{TS,COD}$ [kg/(kg$_{TS}$*d)]	8	0,073	0,023
P_B [l/(m²*h*bar)]	8	84,42	17,44	$B_{TS,BOD5}$ [kg/(kg$_{TS}$*d)]	8	0,036	0,016
pH	8	7,73	0,14	COD:N	8	3,85	1,01
O_2 [mg/l]	8	2,29	0,49	(COD:N)$_G$	8	7,37	1,57
T_{BB} [°C]	8	18,1	1,05	BOD_5:N	8	1,28	0,43
RV	8	7,1	2,38	(BOD_5:N)$_G$	8	4,05	0,96

FM (+ FHM) + MBR (+ PAK) (+ GAK)

Operating parameters	n	average	s	Mud age [d]	15	24,11	2,30
Q_N [l/h]	15	265,73	40,27	TS_{BB} [g/l]	13	14,29	1,99
Q_B [l/h]	15	332,16	50,34	$SV30_{BB}$ [ml/l]	12	1160	164
MF_N [l/(m²*h)]	15	5,43	0,87	$B_{R,COD}$ [kg/(m³*d)]	15	1,95	0,84
MF_B [l/(m²*h)]	15	6,79	1,09	$B_{R,BOD5}$ [kg/(m³*d)]	15	0,72	0,32
P_N [l/(m²*h*bar)]	15	52,80	8,73	$B_{TS,COD}$ [kg/(kg$_{TS}$*d)]	13	0,150	0,101
P_B [l/(m²*h*bar)]	15	65,99	10,91	$B_{TS,BOD5}$ [kg/(kg$_{TS}$*d)]	13	0,055	0,033
pH	15	7,51	0,67	COD:N	15	9,38	3,45
O_2 [mg/l]	15	2,15	0,58	(COD:N)$_G$	15	13,01	3,30
T_{BB} [°C]	12	19,77	0,50	BOD_5:N	15	1,99	1,23
RV	15	8,77	1,65	(BOD_5:N)$_G$	15	4,78	0,96

MBR + PAK (+ GAK)

Operating parameters	n	average	s	Mud age [d]	7	21,03	0,19
Q_N [l/h]	7	296,52	2,72	TS_{BB} [g/l]	7	13,75	1,07
Q_B [l/h]	7	370,65	3,40	$SV30_{BB}$ [ml/l]	7	964	35
MF_N [l/(m²*h)]	7	6,00	0,85	$B_{R,COD}$ [kg/(m³*d)]	7	1,04	0,10
MF_B [l/(m²*h)]	7	7,50	0,92	$B_{R,BOD5}$ [kg/(m³*d)]	7	0,46	0,05
P_N [l/(m²*h*bar)]	7	63,97	9,79	$B_{TS,COD}$ [kg/(kg$_{TS}$*d)]	7	0,071	0,010

P_B [l/(m²*h*bar)]	7	79,96	12,24	$B_{TS,BOD5}$ [kg/(kg_{TS}*d)]	7	0,032	0,005
pH	7	8,01	0,16	COD:N	7	3,09	0,59
O_2 [mg/l]	7	2,16	0,53	(COD:N)_G	7	6,74	0,70
T_{BB} [°C]	7	19,23	0,89	BOD_5:N	7	1,02	0,13
RV	7	5,46	0,05	(BOD_5:N)_G	7	3,93	0,48

Supplementary trial phase T1: MBR + FM

Operating parameters	n	average	s	Mud age [d]	5	21,50	1,58
Q_N [l/h]	5	241,36	8,2	TS_{BB} [g/l]	5	15,81	0,55
Q_B [l/h]	5	301,7	10,20	$SV30_{BB}$ [ml/l]	5	1250	96
MF_N [l/(m²*h)]	5	4,83	0,8	$B_{R,COD}$ [kg/(m³*d)]	5	0,98	0,1
MF_B [l/(m²*h)]	5	6,03	0,9	$B_{R,BOD5}$ [kg/(m³*d)]	5	0,39	0,05
P_N [l/(m²*h*bar)]	5	44,66	4,32	$B_{TS,COD}$ [kg/(kg_{TS}*d)]	5	0,06	0,005
P_B [l/(m²*h*bar)]	5	55,83	5,4	$B_{TS,BOD5}$ [kg/(kg_{TS}*d)]	5	0,02	0,005
pH	5	7,56	0,51	COD:N	5	6,00	1,32
O_2 [mg/l]	5	2,25	0,30	(COD:N)_G	5	11,74	1,64
T_{BB} [°C]	5	16,88	0,62	BOD_5:N	5	2,65	0,93
RV	5	5,81	0,2	(BOD_5:N)_G	5	7,15	0,97

Supplementary trial phase T1: MBR + PAK

Operating parameters	n	average	s	Mud age [d]	5	22,0	1,35
Q_N [l/h]	5	342,41	47,5	TS_{BB} [g/l]	5	13,40	0,51
Q_B [l/h]	5	428,01	59,3	$SV30_{BB}$ [ml/l]	5	1000	0
MF_N [l/(m²*h)]	5	6,84	1,25	$B_{R,COD}$ [kg/(m³*d)]	5	1,04	0,10
MF_B [l/(m²*h)]	5	8,56	1,35	$B_{R,BOD5}$ [kg/(m³*d)]	5	0,62	0,09
P_N [l/(m²*h*bar)]	5	53,97	13,59	$B_{TS,COD}$ [kg/(kg_{TS}*d)]	5	0,07	0,006
P_B [l/(m²*h*bar)]	5	67,46	16,99	$B_{TS,BOD5}$ [kg/(kg_{TS}*d)]	5	0,04	0,006
pH	5	7,67	0,24	COD:N	5	5,37	1,24
O_2 [mg/l]	5	2,28	0,54	(COD:N)_G	5	11,14	1,17
T_{BB} [°C]	5	16,78	0,80	BOD_5:N	5	2,09	0,56
RV	5	4,98	0,67	(BOD_5:N)_G	5	6,64	0,33

References

1. Baumgarten S (2007) Membranbioreaktoren zur industriellen Abwasserreinigung, Dissertation zur Erlangung des akademischen Grades eines Doktors der Ingenieurwissenschaften von der Fakultät für Bauingenieurwesen der Rheinisch-Westfälischen Technischen Hochschule Aachen
2. Dohmann M, Buer T, Vossenkaul K (2002) Stand und weitere Entwicklung membrantechnischer Anlagen im Bereich der Wasserversorgung und Abwasserentsorgung. Vortrag. Büchel, Januar 2002
3. DWA 131 (2016) Arbeitsblatt DWA-A 131- DWA-Regelwerk Bemessung von einstufigen Belebungsanlagen, Hennef

4. DWA-M 22738 (2011) DWA-Merkblatt 227, Membran-Bioreaktor-Verfahren (MBR-Verfahren), DWA Fachausschuss KA 7, Hennef, Entwurf, Stand 27.07.2011
5. FiW eV (2015) SINOWASSER Vorhabensbeschreibung.
6. Hu P (2017) Nachhaltige Klärschlammentsorgungskonzepte für die Stadt Shenyang unter besonderer Berücksichtigung deutscher Erfahrungen, Masterarbeit am Lehrstuhl für Siedlungswasser- und Siedlungsabfallwirtschaft RWTH-Aachen
7. Itokawa H, Thiemig C, Pinnekamp J (2008) Design and Operating Experiences of Municipal MBR in Europe. Water Sci Technol 58(12):2319–2327
8. Jin L, Zhang G, Tian H (2014) Current state of sewage treatment in China. In: Water Research. 66 Jg., S. 85–98
9. Judd S (2011) The MBR Book: Principles and Applications of Membrane Bioreactors for Water and Wastewater Treatment, Butterworth-Heinemann, second edition
10. Krause S, Zimmermann B, Thiemig C (2011) Untersuchungen zum ressourcenschonenden Betrieb von Membranbelebungsanlagen - Optimierungen hinsichtlich Energie- und Chemikalienbedarf, Korrespondenz Abwasser (58) Nr. 9, 2011
11. Li X, Li G (2015) A Review: Pharmaceutical Wastewater Treatment Technology and Research in China. In: Asia-Pacific Energy Equipment Engineering Research Conference. Zhuhai, China, 13–14 Juni 2015
12. MEP (2010) China environmental statistics annual report 2010, Ministry of Environmental Protection of the People's Republic of China
13. MEP (2016) China environmental statistics annual report 2016, Ministry of environmental protection of the People's Republic of China
14. Meteoblue AG (2018) Klima Shenyang 2018. https://www.meteoblue.com/de/wetter/vorhersage/modelclimate/shenyang_china_2034937, zuletzt abgerufen am 23.05.2018
15. Metzger S, Kapp H (2007) Adsorptive Abwasserreinigung zur Verbesserung der Gewässerqualität, Hochschule Biberach, Institut für Geo- und Umwelt, Lehrgebiet und Labor für Siedlungswasserwirtschaft, 15 S.
16. Mitschele M (2013) Untersuchungen zur aeroben Abwasserreinigung ohne biologische Überschußschlammproduktion. (Technik). Hamburg: Diplomarbeiten Agentur
17. NEPG: Website der Northeast Pharmaceutical Group Co., Ltd. http://www.nepharm.com, zuletzt abgerufen am 16.01.2018
18. Pinnekamp J, Benström F, Nahrstedt A, Böhler M, Knopp G, Montag D, Siegerist H (2016) Leistungsfähigkeit granulierter Aktivkohle zur Entfernung organischer Spurenstoffe aus Abläufen kommunaler Kläranlagen. Ein Review halb-und großtechnischer Untersuchungen. Teil 1: Veranlassung, Zielsetzung und Grundlagen. In: Korrespondenz Abwasser. Jg. 63, 2016, Nr. 4, S. 276–289
19. Pinnekamp J, Friedrich H (2006) Membrantechnik für die Abwasserreinigung, Schriftenreihe Siedlungswasser- und Siedlungsabfallwirtschaft NRW, Band 1, 2. aktual. Aufl., FiW-Verlag, Aachen
20. Shan L (2017) Application and Demonstration of sewage treatment and sludge disposal technology in Shenyang. Vortrag des Shenyang Academy of Environmental Sciences, SINOWATER-Symposium 2017, Peking
21. Shenyang EPB (2017) Offizielle Homepage des Environmental Protection Bureau von der Stadt Shenyang. http://www.syepb.gov.cn/data, zuletzt abgerufen am 16.05.2018
22. Tian Z (2017) Upgrading Strategy of North WWTP in Shenyang, Vortrag des Urban Water Environmental Department of CRAES, SINOWATER-Symposium 2017, Peking
23. Xiang L (2017) Treatment Technology of Pharmaceutical Wastewater. Vortrag des Chinese Research Academy of Environmental Sciences, SINOWATER-Symposium 2017, Peking
24. Xiao S, Zhang G (2011) Treatment of Berberine Pharmaceutical Wastewater Containing Copper by Bipolar Electrochemical Process. In: Environmental Engineering Technology. Jg. 1, 2011, Nr. 1, S. 295–299
25. Yun Y (2018) Upgrading of wastewater treatment plants in Germany and in China - progress and trend, a contribution to the closing Symposium Integrated Lake Basin Management Dianchi, Vortrag des Forschungsinstitut für Wasser- und Abfallwirtschaft an der RWTH Aachen (FiW) e.V., SINOWATER-Abschlusssymposium 2018, Kunming

26. Zhang P (2006) 沈阳铁西工业区改造的制度与文化因素 Human Geograph, vol. 21, No. 2, 04.2006
27. Zeng P, Du J, Liu Y, Gao H, Liu R, Song Y (2015) Pharmaceutical wastewater treatment in China and the world. A bibliometric analysis of research output during 1990–2013. Res. Rev. J. Pharmaceutical Qual. Assurance. Jg. 1, 2015, Nr. 1, S. 30–37

Chapter 5
The Planning, Management and Decision Support Systems of Kunming's Urban Drainage System

Hailing Wang, Liu Dan Yu, Paul Wermter, and Florian Rankenhohn

5.1 Overview

5.1.1 Chinese Policy Background

To enhance the effectiveness of prevention and control of water pollution, a series of management requirements and technical targets are proposed in national and industry related policy documents. For example, the action plan for water pollution prevention and control demands for the rapid transformation from the existing combined drainage system to separated sewage system. If it is difficult to transform a given system of combined drainage, the measures such as intercepting the run-off, regulating and storing, and controlling should be taken into consideration. It is necessary for each city and region to determine the transformation mode of drainage system according to the demand of local water management.

H. Wang (✉)
Kunming Dianchi Investment Co., Ltd, Yunnan 650228, P. R. China

L. D. Yu
Guangzhou Municipal Engineering Design and Research Institute Co., Ltd., 348 Huanshidong Road, Guangzhou 510160, China

P. Wermter · F. Rankenhohn
Research Institute for Water and Waste Management at RWTH Aachen (FiW), Kackertstr. 15-17, 52056 Aachen, Germany

F. Rankenhohn
Department Resource Efficient Wastewater Technologies, University of Kaiserslautern, Paul-Ehrlich Street 14, 67663 Kaiserslautern, Germany

P. Wermter
Ministry of Climate Protection, Environment, Energy and Mobility of Rhineland-Palatinate, Kaiser-Friedrich-Street 1, 55116 Mainz, Germany

© The Author(s) 2022   169
M. Dohmann et al. (eds.), *Chinese Water Systems*, Terrestrial Environmental Sciences,
https://doi.org/10.1007/978-3-030-80234-9_5

MHURC (2015)[1] [1] requests that the drainage water quality of lakes must not be lower than the surface water standard class IV. Except for the period of rainfall, the combined sewer should not directly discharge into the water body. The Technical guide lines for urban wastewater drainage requests that in areas where drainage pipes are laid below the groundwater level, the inlet water's chemical oxygen demand (CODcr) concentration of municipal wastewater treatment plants (WWTP) cannot be lower than 260 mg/L in dry days, or increase by 20% per year on the basis of the existing water quality concentration. The '13th five-year plan' for the construction of urban wastewater treatment and recycling facilities issued by the national development and reform commission and the ministry of housing and urban-rural development, demands to speed up the resolution of the uneven layout of wastewater treatment facilities and the implementation of stricter discharge standards in areas with serious water pollution.

The local government responded positively to the above policies. The implementation plan of the three-year critical action for the protection and governance of Dianchi lake (2018–2020) demanded that in 2018, Dianchi Caohai's water quality should reach class IV (compare Table 5.1). Dianchi's water quality throughout the year reached classes V, but was better in the dry season with class IV. In 2020 Dianchi Caohai and Waihai water quality should achieve class IV. In addition, the Chief River Commander's Order signed and issued by Kunming municipal party committee secretary in March 2018, demands to

- complete the construction of diversion and retention facilities for the channel and tributaries
- prevent the overflow when the initial rainfall of the channel and tributaries is 7–10 mm/d
- effectively reduce the pollution load of the water from the combined sewer flowing into lakes and rivers.

Furthermore, in order to effectively control the eutrophication of Dianchi water body, authorities ask to strengthen the control of discharge of major water contaminants and the so-called recycling of wastewater from the urban WWTP or re-use of urban wastewater. Therefore, Kunming compiled the main water contaminants discharge limit values for urban WWTP in Kunming (trial). The discharge limits of major water contaminants under normal operation mode of urban WWTP in Kunming administrative region and the limits of discharge of major water contaminants under rainy season operation mode of overflow sewage in rainy season were stipulated and stricter discharge standards were implemented to further reduce pollution in Yunnan.

---

[1] City office letter [2004] no. 153 issued by the Ministry of Housing and Urban-Rural Development of the People's Republic of China (MOHURD).

Table 5.1 Discharge limit of major water contaminants in Kunming urban WWTP (trial)

Contaminants	Grade A	Grade B	Grade C	Grade D	Grade E
BOD5	4	6	10	10	30
COD	20	30	40	50	70
NH4-N	1.0 (1.5)	1.5 (3)	3 (5)	5 (8)	15 (20)
TN	5 (8)	10 (15)	15	15	30
TP	0.05	0.3	0.4	0.5	2

Values in brackets are valid from December 1 (inclusive) to March 1 (exclusive).

5.1.2 Current Status and Problem Analysis

Since the 11th five-year plan, a large number of water environment treatment projects have been built in the Dianchi basin, especially in the main urban area of Kunming. The administration of the main urban area of Kunming has built 15 WWTP (two are still under construction, compare Fig. 5.1). The total service area of the main urban sewage system is 518.8 km². The wastewater treatment capacity reaches nearly 2 million cubic meters per day, and the effluent quality is better than grade A. 19 storm water overflow tanks have been built (two of which are still under construction). In addition, 5,722 km of municipal network for drainage and 96 km of a circular sewerage system for collecting sewage around the lake have been built. Together, these components are forming an integral urban drainage system (compare Table 5.2).

On the one hand, the construction and operation of these projects have improved the collection and treatment of the domestic sewage in the dry season, and effectively reduced the overflow frequency of combined sewers in the rainy season. It helped reducing the pollution load, protecting the river channel and ensuring the safety of water quality which flow into the Dianchi lake. The Evaluation on the implementation of the 12th five-year plan for Dianchi lake governance compiled by the Chinese academy of engineering affirmed the achievements of Dianchi lake governance in recent years. The water quality in Dianchi has been stabilized and continuously improved. The pollution load from point sources has been significantly reduced.

On the other hand, because a transformation from combined to a separated sewer system is a difficult and complex task, the urban combined sewer system will still exist for several years. The relationship between runoff and performance of the drainage system has become increasingly complex. It is difficult to control the terminal drainage facilities that serve for both: the drainage of pollution in the dry season and the flood discharge in the rainy season (compare Fig. 5.2). This complex situation causes overflow discharges with negative effects for the receiving water bodies. Not only that, but also the treatment load between different WWTP is extremely unbalanced. There is yet a lack of data for the upstream pump stations of the receiving WWTP. The setting and timing of the stored capacity and discharged capacity of controlled retention facility lacks a sound data basis, and it is not effective

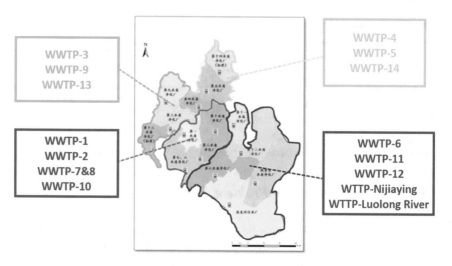

Fig. 5.1 WWTP within separate drainage areas in Kunming

Table 5.2 Overview of WWTP in Kunming city centre

Drainage Systems of...	Waste Water Treatment Plant	Catchment Area (m²)	Planned Population (10⁴ m³)	Planned Population Density (10⁴ per sq km)	Mean Inflow (10⁴ m³)	Average Daily Capacity (10⁴ m³)	Maximum Capacity in in Wet Yeather Condition (10⁴ m³)
Western Catchment Area	WWTP No. 3	27,68	45,95	1,66	14,6	23,42	26,12
	WWTP No. 9	25,72	39,87	1,55	12,7	6,47	7,95
	WWTP No. 13 (Under construction)	20,06	21,32	1,06	6,8	N/A	N/A
Northern Catchment Area	WWTP No. 4	12,93	27,54	2,12	8,7	4,66	4,91
	WWTP No. 5	33,14	67,55	2,04	21,5	23,56	25,09
	WWTP No. 14	23,31	22	0,94	7	N/A	N/A
Southern Catchment Area	WWTP No. 1	17,95	36,95	2,06	11,7	13,85	15,67
	WWTP No. 2	24,21	54,03	2,23	17,2	11,73	12,47
	WWTP No. 7 & 8	37,11	36,74	0,99	11,7	32,61	36,39
	WWTP No. 10	20,32	46,7	2,3	14,8	9,55	10,58
South-Eastern Catchment Area	WWTP No. 6	39,1	63,32	1,62	20,1	14,1	17,6
	WWTP No. 11 (Under testing)	26,31	21,77	0,83	6,9	1,74	2,88
	WWTP No. 12	47,15	26,4	0,56	8,4	No data	No data
	Nijiaying WWTP	66,13	39,02	0,59	12,4	No data	No data
	Luolonghe WWTP	97,68	60,56	0,62	19,2	6,18	7,21
Total (or average)		518,8	609,63	1.18 (average)	193,7		

to deliver all the treated sewage to the plant. In addition, the operational performance of completed governance projects is lacking of data, too, and the project operation is mainly focused on node optimization. With the increasing requirements for water quality protection of Dianchi, the consciousness for a dedicated water environment management system is also increasing. Therefore, new governance problems have posed new questions:

- How to reduce emissions by managing and controlling the system by using synergies of existing facilities?
- How to make decide on a scientific basis, whether to build new facilities or restore existing facilities

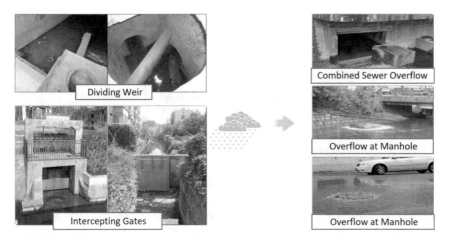

Fig. 5.2 Drainage systems facilities and examples for severe overflows

- How to improve the management of engineering measures and non-engineering measures in the governance of Dianchi.

5.1.3   Methods

In conclusion, the requirements for water environmental management of the "National Policy" are significantly increased. Therefore, the determination and the pressure to carry out Dianchi lake management are gradually increased. In the face of the problems of the Kunming drainage system and related treatment projects, effective storm water management measures are as follows:

- Supporting the objective of water quality improvement, the authorities carry out precision pollution control within the Dianchi basin
- Promoting the objective of capacity improvement, the authorities will use engineering governance decision-making;
- Stimulating the objective of effectiveness improvement, the authorities will implement a united regulating system.

The "source - network - station - plant - river - lake" integrated linkage management mechanism and technical system will be established for the total system performance maximization. It aims to combine, the overall planning of the rainfall collection, treatment and reuse facilities. This integrated approach should strengthen the interaction between the effectiveness of facility system and the quality of water environment. The integration will hugely impact on planning and design, operation management, optimization control and decision-making of Kunming urban drainage system. Finally, let the project decision and system operation of Dianchi lake management become scientific and intelligent.

5.2 Managerial Decision Support System of Urban Drainage System

In order to scientifically and systematically govern Dianchi, data fusion management is required. It is necessary to rely on information means to assist the planning of Dianchi lake governance project, the scientific management of drainage facilities and the efficient operation. By establishing a comprehensive database with a standardized format, complete content and the latest updates, the data management and monitoring of massive drainage facilities can be realized. It serves to fully understand and get a clear picture of the current situation of drainage facilities. The management is aiming at establishing a systematic simulation decision-making tools and multi-level effectiveness evaluation index system. It should be combined with the monitoring data and multiple scenario simulation analysis and comparison. This system provides support for the decision-making and evaluation of Dianchi lake treatment project and the regulating control of drainage facilities. It not only enhances the capacity of decision support for major projects, but also optimize operational efficiency and effectiveness of existing facilities.

5.2.1 Data System

Data is the basis of decision-making. The goal is to realize the standardization and unification of a large amount of planning data, structural data and engineering operations and scheduling management data (compare Fig. 5.3). There is a need for data on drainage catchment areas, drainage systems, combined overflow basins, WWTP and other parts of the drainage system. Authorities will rely on a unified data centre to carry out data management. The unified data centre should make it connectable, shareable, and serviceable for decision-making. It should support data update and improvement and long-term accumulation in the process of construction and renovation of drainage facilities. Finally, it may provide long-term support for the evaluation of the application data, optimizing the design scheme of drainage facilities planning and regulating operation of engineering.

Firstly, the data standards need to be unified, and the data structure of the GIS data related to drainage of Dianchi lake needs to be harmonised and agreed respectively. Secondly, a unified management platform of database for drainage facilities should be developed based on data standards. Based on GIS technology, the platform realizes the display, query and edit of Kunming drainage facilities data, and maintains the complex network topological relationship of facilities dynamically. It can be

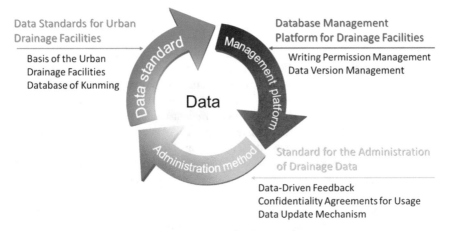

Fig. 5.3 The iterative update mechanism for the basic data of Kunming drainage facilities

used for facility connectivity and upstream and downstream analysis. Furthermore, it can be used as the supervision software for drainage GIS data rapid analysis and measuring projects. The uniqueness and effectiveness of the database are guaranteed by using unified data management software. Finally, a unified data management method should be formulated to clarify the feedback update mechanism and continuous update mechanism for the data, as to improve the refresh rate of data. Through appropriate data confidentiality mechanisms, the data security can be guaranteed and the abuse of urban infrastructure data can be prevented.

Through the construction of the unified data system, the drainage data standards, software tools and management mechanism of Dianchi basin are standardized. The existing resources can be fully shared to improve the investment efficiency of data system construction and the constant and frequent use of data.

The collection and transmission system should be constructed for all kinds of Dianchi lake operation data like:

- water quality and quantity data for key drains,
- water quality and quantity data of input and output water in storage pool
- key operation parameters, water quality and quantity data of Niulan River,
- the water quality and quantity data of tail water,
- the water flow and quality of the western traffic tunnel,
- water quality data of the main canal which intercepts sludge around the lake, etc.

By the means of data standardization in the projects, the effective accumulation and quick access of project-related data can be realized to ensure the long-term operation of the data centre.

5.2.2 On-line Monitoring System

According to the project decision-making requirements of monitoring data of local drainage system in Kunming, the advanced "Internet of things" technology is adopted. It will help to carry out the whole monitoring process of discharge sources, branch manholes and main manholes, pumping stations, storage facilities and WWTP of Kunming drainage system. Furthermore, meteorological observation data, river and lake water quality monitoring data and river hydrological monitoring data are processed and added to the system. The responsible institutions for the data are Kunming meteorological bureau, environmental protection bureau and hydrological and water resources bureau. The Dianchi lake treatment information collection system will be established with reasonable layout, clear level, complete functions, soft and hard integration and long-term effectiveness. Finally, the integrative comprehensive perception of "source-network-station-WWTP-river-lake" should be realized.

Based on the current evaluation, the online monitoring and early warning system for urban drainage network was established and operated in a long-term. It follows and holistic approach for the monitoring of the drainage network. It ensures continuous, accurate and updated monitoring data and helps to provide a reliable decision-basis for drainage engineering. Ensuring long-term stability, a standardised supervision mechanism is established. It can monitor the operation status and calculate risks for drainage facilities dynamically, quickly track and warn the drainage system management department. By collecting the long-term drainage operation data, it can be used to identify necessary maintenance and helps to derive operation rules for drainage facilities. It assesses and diagnoses the current operation status of drainage facilities, supports the decision of major projects, and assesses the implementation effects of sponge city, black smelly water, drains and other related projects quantitatively.

During the construction process of the online monitoring system, the following monitoring requirements are mainly considered:

(1) Collection and accumulation of long-term monitoring data of the status-quo drainage system. Relevant facilities should continuously be monitored and tracked to accumulate data, so as to form localized technical criterions and provide experience and guidance for similar urban drainage pipe network construction and transformation. It cannot only improve the scientific and effective management level of the pipe network, but also improve the operation efficiency of the drainage system. It also supports the modification of the parameters and verification of the calibration of the mathematical model. And it also provides the basis for real-time warning forecasting and dynamic assessment.

(2) Programming the scale of reconstruction design scheme and determining the reference basis of the planning and reconstruction design scheme. Based on accurate data and the mathematical model it provides the reference for the objective design and determination of pump station, sewage plant and other

facilities, and objective evaluate the advantages and disadvantages of existing drainage system. According to the existing problems, propose the improvement for the planning scheme, and conduct scientific and systematic optimization.

(3) With the continuous improvement of drainage pipe network construction, advanced monitoring and modelling is used to support the scientific decision-making and performance evaluation of facilities in accordance with relevant technical criterions and under the guidance of scientific concepts.

5.2.3 Model Simulation System

In order to comprehensively and objectively evaluate the operation effect of Dianchi treatment facilities, it is necessary to construct a comprehensive simulation system which integrates the source-network-station-WWTP-river-lake system, to realize the simulation analysis and evaluation of the system.

Firstly, a dynamic model for pollution flux in the land and water areas of Dianchi Lake should be established. It simulates the main processes of rainfall, runoff, nonpoint source pollution-point source import - pollutant flux and transfer - surface emissions and the entrance of runoff and pollutant loads, into Dianchi through tributaries. Then the data of sewage and pollution load, treatment reduction and water inflow for each sub-basin were obtained.

Secondly, a 3D-model of water quality and algae for Dianchi needs to be established to simulate how pollutants are transported and transformed within the waterbody. It is designed to analyse the relation between amount of water in tributaries and the water quality of contaminated waterbody of the Dianchi. Combining a land-water management model, helps to better understand the contribution of the tributaries on the water quality of Lake Dianchi. The analysis of the major factors which are impacting the water quality and the duration of water not meeting the standard of the requirements, helps identifying the water reaches that would meet the requirements. It would also help to find as the impacted watercourses, estimating the control value of contaminants in each estuary and the targeted value to reduce to.

At the same time, it is necessary to model the drainage system. The operating rules of the drainage system on dry days and rainy days were simulated and analysed with the water quantity and water quality constraint conditions of the given drainage system. Make a systematic evaluation of the design of drainage pipe networks, retention facilities and WWTP, and analyse and study a variety of optimal operation scheduling strategies such as integrated design of multi-pump stations, trans-regional control, and the integrated design of pipe network and WWTP, retention facility and pump station, etc. The mathematical model is used to carry out annual simulation analysis and evaluation on the implementation effect of various optimized operation and design strategies and analyse the response relationship between drainage facilities and water environmental protection.

Finally, in order to maintain data quality and also keep up with the current data need, regular upgrade and maintenance are needed for the database and models based on land use change, management change, change of layout of the treatment plants and the change of rainfall, etc. so that decision making would become easier.

5.2.4 Performance Evaluation System

The evaluation objectives, decision-making objectives and specific evaluation indexes of the system are determined hierarchically (compare Table 5.3). It supports the scientific decision making of Dianchi management by evaluating the situation and management scenarios.

Models are used to simulate the ongoing operation of the drainage system and to draw the water balance relationship diagram of the study area. The operating condition of the drainage system in the area is analysed. Analysis parameters are in particular core indices such as the overflow frequency, the annual overflow rate of the retention basin/catchment area, the annual overflow frequency of the controlled overflow structure, the retention facility, which is coordinated with the annual wastewater treatment quantity and pollution load of the WWTP. The result is the annual evaluation report and conclusions for problem diagnosis. Existing problems in the current operation scheme are pointed out and the inconsistencies of the existing drainage facilities are clarified.

In the decision-making process of the Dianchi Basin, major projects affecting the drainage system should be based on the above-mentioned model. The decision support should address site selection and scale determination of the WWTP, the site selection and scale determination of the retention facility, the construction of major sewers, the reconstruction or new construction of large pumping stations, in particular, projects that have a large impact on the operation of upstream and downstream related drainage facilities, complicated and variable boundary conditions, and inter-connected relationship upstream of the facility. We need to collect necessary monitoring data, and determine and validate model parameters. The mathematical model is used to simulate and analyse the system response under various working conditions, multiple scenarios and multiple different planning schemes. It is used to quantitatively evaluate the impact of key points, and support the planning and control of related major projects. After the project is completed, the effect evaluation should be based on the measured data, and the model should be used to optimize and improve the operation and design strategy, improve the investment operation efficiency of major projects, and maximize the efficiency of project investment.

Table 5.3 Parameters of performance evaluation

Evaluation of …	Objective of decision	Specific evaluation index (Gross/arid day/rainy day)
Receiving water	Evaluating the overall pollution control effect of the basin dynamically, to decide the focus of systematic pollution control projects and clarify the key pollution control catchment area.	Control the total pollution load before discharging into the river Target reduction of the regional pollution load Water quality achieved rate of key section
Service zone or Catchment area	Making decisions on the overall project scale and related control index of the area drainage collection and treatment system, in order to ensure that the river water quality does not degrade from upstream to downstream.	Regional pollution load Pollution load collected and treated by regional drainage system Pollution load reduction rate Total pollution load discharged by the regional drainage system Overflow pollution load of storm water overflow discharge Control rate of total annual runoff / Annual overflow frequency, and annual control rainfall in millimetres of overflow
Engineering facilities – drainage system	Making decisions on drainage system construction, reconstruction, maintenance, strategic dispatch or Unicom	Overloading rate of drainage system Collection rate of drainage system Achieved rate and increase rate of influent COD concentration of the drainage system.
Engineering facilities – retention facilities	Making decisions on retention facilities' scale, key parameters of running schedule	Annual reduced overflow frequency of retention facilities Annual wastewater treatment capacity and pollution load of WWTP matched by retention facilities
Engineering facilities - WWTP	Making decisions on WWTP scale, processing capacity and running model	Pollution load reduction of WWTP Total pollution load of tail water discharged by WWTP

5.3 Progress in Practical Application

5.3.1 Data Construction and Application

5.3.1.1 Data Standard Establishment

Digitization of the drainage system is a complicated and systematic project, characterised by a long construction period. The accurate and effective collection of drainage data is the key to the effectiveness of digitization of the drainage system. During the 12th Five-Year Plan period, Kunming City carried out the construction of drainage facilities such as retention facilities and pumping stations. Combined with the needs of engineering and operation management, several digitization projects of the drainage system projects have been launched. Due to lack of standard requirements for drainage facility data, standards of drainage monitoring equipment and monitoring strategies during the design and construction process, some drainage data cannot be collected accurately. This is leading to a lack of key data for drainage facility management. It can ultimately affect the effectiveness of drainage digitization projects. The technical level of drainage facility design and management is effectively improved based on the lesson learned from the digitization of existing drainage systems. Therefore, we need to establish relevant standards, standardize and guide the design and construction process of drainage facilities and reserve the integrated data interface for post-information system construction.

Based on the relevant digitization experience of the Kunming Dianchi Investment company, the Technical Standards for Data Management of Urban Drainage Facilities in Yunnan Province (DBJ53/T-93-2018) [2] were established. This realized the regional expansion of national standards and the establishment of a harmonised data format, complete content and dynamic update for an urban drainage facilities basic database. At the same time, the Technical Standard for Online Data Collection of Urban Drainage of Yunnan Province (DBJ53/T-9 4 -2018) [3] was established. This set standards for online data collection and transmission of drainage facilities, harmonised the list of key information data and technical standards for drainage facilities data such as retention facilities, pumping stations and pipe networks. It also ensures the timeliness and authenticity of the dynamic data of drainage facilities. The preparation and implementation of the above data and collection standards ensure the uniformity of the GIS data and dynamic monitoring data of the drainage facilities at the source, and provides reference for the compilation of other relevant standards of the Dianchi governance information system.

The next steps will be based on the technical needs of digitization of Dianchi governance, combined with the promotion and development of related projects, summarizing and refining, compiling corresponding technical standards, standardizing technical work and related data, thus providing a complete standard system for digitization of Dianchi governance.

5.3.1.2   Data Platform Construction

The network topology structure follows the principle "one centre, two sub-stations, several field control points". The investment company is the host for the data centre, and the drainage company and Dianchi water service were used as the extension sub-station to realize the on-demand calls to data resources and the application (compare Fig. 5.4). It is integrated with nearly 70 core hardware devices such as necessary servers, workstations, disk arrays, switches, and intrusion prevention devices, and deployed with more than 60 sets of core basic software such as database, ArcGIS, configuration software, PLC programming software, watershed drainage model software, network security management software, antivirus, network video software and so on. These field control points provide reliable guarantee for daily network operation and maintenance, data storage and backup, video online monitoring and information system operation and development, and provide basic guarantee for the stable operation of the company's information systems.

The systems launched are:

- the development practice of the Kunming main city pipe network information system,
- Kunming main city retention facility information system,
- Kunming lake East coast information system,
- the Western Catchment Operating System and
- a series of application software.

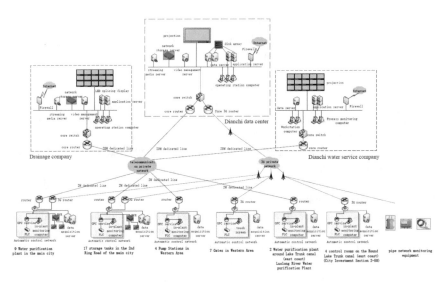

Fig. 5.4 Network topology following the principle "one centre, two sub-stations, several field control points"

The systems are based on the above-mentioned basic software and hardware and network platform (see picture above), combined with the specific management needs of various periods and various engineering projects, through the customized development method. It provides a corresponding business subsystem for the asset management, monitoring and retention facility monitoring and management of the Kunming drainage network, and drainage system operating analysis.

Based on the experience, they developed information management function modules for some regions and some facilities. The modules provide a functional foundation and valuable experience for the next step of digitization.

However, in general, the existing related application software systems are not strong overall. The functions are strongly limited, the operation methods are not user friendly enough, some functions are isolated, the operation and maintenance are not in place and a complete set of Kunming drainage information application software systems have not yet been formed. In the next step, the development needs to be based on the overall requirements of drainage management, unified data standards and system architecture to plan the software application functions. It also integrates existing related modules, information and functions, providing an integrated visual management decision window of drainage planning management in Kunming.

5.3.1.3   Data Feedback and Application

With the rapid development of Kunming, the urban drainage facilities are constantly upgraded. Based on years of experience in drainage facility data management, Kunming Dianchi Investment has proposed a management mechanism for data feedback and update. The data feedback and update workflow include data collection and processing, on-site data monitoring, data management and storage, data applications and feedback as shown in Fig. 5.5.

Before the urban drainage facilities data is updated, new and reconstructed drainage blueprints within the data detection scope of this stage are collected, and the data problems reported by the data use unit are formed to a fault detection map. The effective development of the fault detection map clarifies the scope of data measurement and improves the working efficiency of the field measurement unit. At the same time, through the spatial superposition analysis with the existing drainage facility data, a repeated mapping of the drainage facility data is avoided. This could save money for data collection.

At present, most urban drainage facilities data detection does produce unclear relationship between mapping results and existing data topologies. In the management mechanism of data feedback and update the connection requirements of the data detection results are clarified. It provides evidence for establishing clear data on urban drainage facilities. The main connections are shown in Fig. 5.6.

In 2015, in the process of establishing the drainage system water model in the southern part of the main city, the data of the 101.2 km drainage pipes was manually supplemented (compare Fig. 5.7). This was done due to the incomplete data at

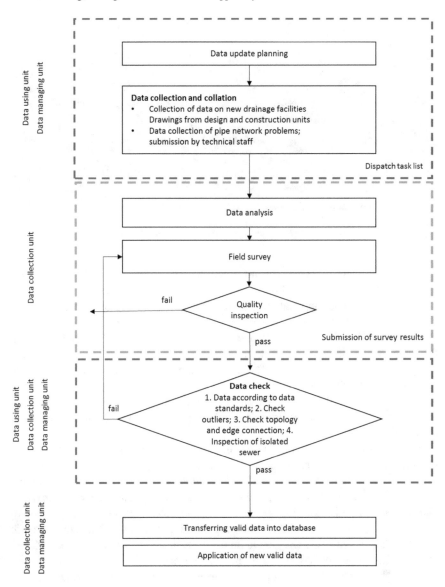

Fig. 5.5   Data feedback update workflow

the time, which caused the uncertainty of the simulation results and affected scientific decision making. However, through the implementation of the Kunming main city data measurement project (Phase I), the measurement group supplemented and improved the 174.4 km drainage pipe data, and the length of the pipeline coincident with the manually added pipeline was 84.26 km, and new drainage system data of 90.14 km was added. The data complements the integrity of the urban drainage

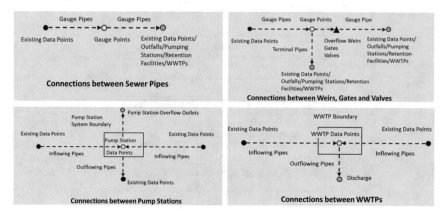

Fig. 5.6 Integration of drainage facilities in the data management system

facility database, provides the basic information for the model analysis of the drainage system, and improves the credibility of the model analysis results.

In the data feedback application process, the tracking and improving of the existing drainage system is supported as well. For example, after verifying the misaligned of the QIAN WEI West Road sewage on site (compare Fig. 5.8), combined with the hydraulic model analysis of the area and the actual flooded water situation, a comprehensive plan for the transformation of the displaced manholes and for the prevention of floods is proposed.

With the help of the model, the impact of the node is systematically evaluated. The operation of the pipeline after the transformation of the misaligned manhole is shown in Fig. 5.9.

After the implementation of the misaligned manhole reconstruction project, the pipeline operation status has improved significantly, and the pipeline overflow capability has been improved. The modified manhole can match the demand for the overflow capacity of the sewage pipeline in the development and construction of the Qianwei West Road area. This case of finding suspected data and making improvements is also an indication of the importance of data verification feedback. Through doubtful data, it can help managers to improve the problem of drainage network and improve the operational efficiency of the overall system.

Through the continuous data updates of several projects, a unified drainage facility data update software platform has been initially developed. It is based on unified data and a complete GIS database of the city's drainage facilities, and has more than 4,100 km of drainage facilities data.

At the same time, a dynamic data acquisition system is connected to relevant dynamic monitoring data of the facilities. It consists of objects such as main sewage plant, retention facility, pump station and pipe key nodes in the main city. At the same time, a dynamic update mechanism for data has been established. Through the method of annual measurement and verification, it is possible to continuously update and improve the basic data of drainage facilities such as drainage pipes,

━━━━   Initially added data

────   Data added after field
         measurment

Fig. 5.7 Overlay analysis of data updated by measure and by manual addition in the Southern Catchment

ditches, rivers, sluice gates, valves, discharge ports, drainage pump stations, WWTP, etc. It can continuously improve the integrity and quality of current drainage data.

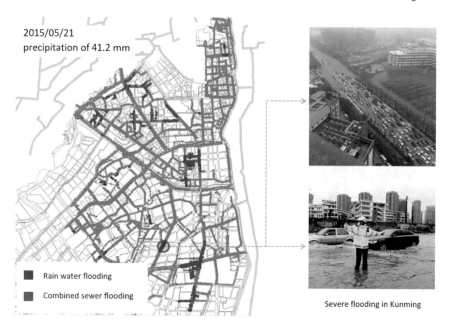

Fig. 5.8 QIAN WEI West Road flooding situation

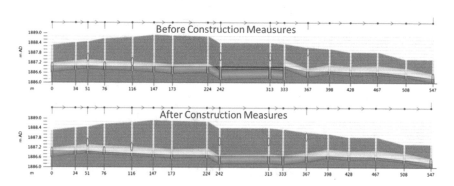

Fig. 5.9 Operation of misaligned manhole before and after project implementation

5.3.2 Online Monitoring Application

5.3.2.1 Monitoring Network Construction

The monitoring network design is based on the needs of related projects. It aims to determine reasonable monitoring points. The arrangement of monitoring points should be closely linked with the purpose of acquisition and application of monitoring data (compare Table 5.4). The purpose of monitoring should not only meet short-

Table 5.4 Monitoring stations in the Kunming drainage system

Project	Types	Quantity	Subtotal	Total
Monitoring of the circular sewer around the lake	Flowmeter	8	33	135
First stage of coordinated and efficient project	Flowmeter	25		
Northern catchment water level monitoring	Water level gauge	6	101	
Southern catchment water level monitoring	Water level gauge	6		
Western level water level monitoring	Water level gauge	19		
Drainage company drainage online water level monitoring	Water level gauge	70		
Panlong River main stream water quantity automatic monitoring station	Flow monitoring station	1	1	

term assessment and verification needs, but also focus on long-term data demand on planning and management of drainage system. Based on the fully understanding of the local drainage network, the river, land use type, urban water accumulation point, and the status quo of the project improvement project, the location of monitoring site is arranged (compare Fig. 5.10).

In the monitoring of the drainage network, not all monitoring points are installed permanently. Instead, we consider temporary measurements to reduce monitoring costs. There are three types of data monitoring services: single sampling, short-term temporary monitoring, and long-term temporary monitoring. It is necessary to combine on-site investigation and on-site inspection to optimize the selection of monitoring methods.

(1)   Conduct field investigations on suspected areas, configure portable liquid level flow meter to quickly monitor instantaneous flow rate, and select and test the sample with serious problem according to the size of the drainage and the pollution status.

(2)   Discharge orifices or manholes with larger instantaneous monitoring flow or larger suspected problems: Quickly install related online monitoring equipment for online monitoring of short-term dry season flow, and obtain continuous flow monitoring data for at least 7 days to ensure the validity of monitoring data and collect monitoring data online.

(3)   For areas with outstanding short-term temporary monitoring problems or areas suspected of mixed rain and sewage, install relevant online monitoring equipment for on-line monitoring of long-term temporary flow. The continuous flow monitoring data cover at least 6 typical field rainfall events, each monitoring point lasts for 2 to 3 months during the rainy season to ensure the validity of

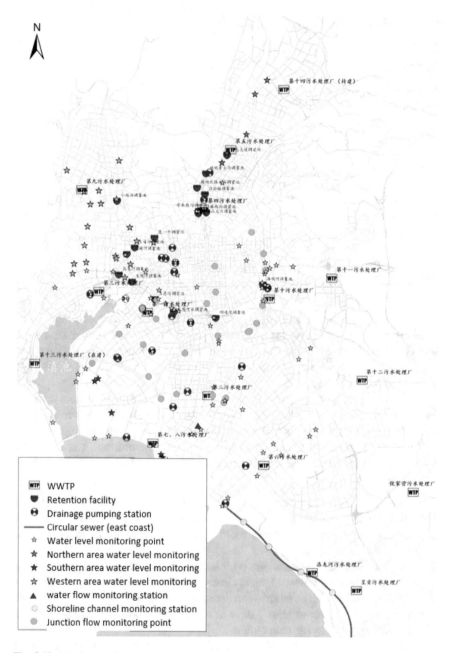

Fig. 5.10 Monitoring facilities

monitoring data and collect monitoring data online. This allows for the selection of a reasonably applicable monitoring method based on the specific needs of the project.

In the construction process of the monitoring system, it is necessary to pay more attention to the maintenance during the operation. This will ensure the quality of the data and ensure the successful operation of the equipment, thus providing effective and timely monitoring data for drainage management.

5.3.2.2   Water Quantity Analysis and Decision in the Northern Catchment

At present, the use of monitoring data for decision support in Kunming is mainly reflected in the long-term monitoring and quantitative evaluation of the drainage water system in the dry season and the rainy season. Based on a water balance analysis, the actual amount of water in the service area of the sewage treatment plant is analysed. This provides data support for the potential tapping and reconstruction of the district's sewage plant. In addition, based on the water quality balance analysis, the inflow and infiltration of the area, the pollution by storm water discharge, and the implementation of diversion and transformation, treatment and separation projects are studied.

A more representative implementation of monitoring-based decision support applications is the rapid monitoring and evaluation of dry season traffic in the Northern Catchment in 2017 (compare Fig. 5.11). The purpose of the project is to obtain the real-time monitoring data of the dry season sewage discharge in the Northern Catchment of the dry season. This is based on the one-month flow monitoring and water quality test in the dry season. A second purpose is to effectively analyse and evaluate the distribution of sewage flow in the dry days of the area, the amount of sewage in the service area of the WWTP and the pollution load. In

Fig. 5.11 Implementation of monitoring equipment

response to the problem of inflow and infiltration of external water, detailed flow monitoring and necessary water sampling and laboratory analysis are carried out to quantitatively determine the inflow and infiltration of external water, and identify the reason for the low influent concentration of the WWTP during the dry season. This provides monitoring data basis for the current situation analysis of drainage network, planning and scale analysis of WWTP, and provide scientific support for the improvement of the drainage network system and the water quantity regulation of the WWTP and scale assessment of the new WWTP.

In order to understand the amount of sewage in the catchment area of the fourth and fifth WWTP, 19 continuous monitoring points and 3 temporary monitoring points are set up in the upstream sewers and upstream of the fourth and fifth plants respectively. They represent the connection between the fourth and fifth WWTP and the sewers in the area with the Second Ring North Road as a boundary.

Based on the statistical analysis of the monitoring data the following key conclusions were drawn: (1) the total amount of pollution generated in the catchment area of the fourth and fifth WWTP are basically in equilibrium with the statistical processing volume. Under the premise that the WWTP starts the first-level intensive treatment, the fourth and fifth WWTPs can basically eliminate the amount of sewage produced by the Northern Catchment; (2) Based on the measured discharge and water quality data, it is estimated that the amount of external water in the fourth WWTP accounts for 64% of the total treated volume, reaching 83,000 m^3. The amount of external water in the fifth WWTP accounted for 77% of the treated volume, reaching 150,000 m^3. In the monitoring area, external water accounted for more than 70% of the total water. (3) The 14th WWTP has been analysed. The increasing population and thus, the significant increase in sewage, characterises the development of the northern catchment area. Therefore, the long-term planning dimension of the fourteenth WWTP is at about 200,000 m^3.

The core functionality of the monitoring and assessment project for the dry season in the Northern Catchment area is to provide strong data-based decisions support for the design and operation of the new fourteenth WWTP in the northern catchment area. Moreover, it has the character of a pilot project for the new WWTP in the southern and western catchment areas.

5.3.2.3 Equilibrium Analysis of Water Volume for the Eastern and Southern Catchment Areas

To support the water balance analysis in the eastern and southern catchment areas, it is necessary, to effectively collect data about the drainage system's status in the dry and rainy season. This consists of analysis of the distribution of sewage in dry and rainy days, the discharge of sewage generated by the districts, the impact of rainfall on the sewage system, and the dimensions of the sewage treatment plants in the area. The second and seventh WWTP are operating at their design maximum, even at dry weather days. The tenth WWTP has not reached the design maximum, but has problems with a low efficiency. By the implementation of the flow measurement service, high-quality monitoring data can show the amount of sewage in dry and rainy days in the catchment area. It analyses operation risks or changes and accurately assesses the water distribution and water balance of the sewage system during the dry season. Thus, the flow measurement is set up in order to analyse and diagnose the inflow and infiltration of dry and rainy days in the drainage network.

According to the topological relationship of the four WWTPs and drainage facilities in the eastern and southern catchment areas of the main city, 25 discharge monitoring points were set up in the WWTPs incoming sewers.

The monitoring system consists of

- 10 monitoring points in the service area of the main city east area,
- 15 monitoring points are set up in the service area of the southern part of Kunming main city.

  - Among them, the first WWTP system sets 5 discharge monitoring points;
  - the seventh and eighth WWTP system sets 5 discharge monitoring points;
  - the second WWTP system sets 10 discharge monitoring points;
  - the tenth WWTP system sets 5 discharge monitoring points.

- At the same time, a rain gauge was distributed in the first WWTP, the second WWTP and the tenth WWTP.

Three rain gauges are used to obtain rainfall data to identify rainfall events, to analyse the water volume in the dry and rainy days, and to verify the drainage system model in dry days and rainy days. The specific location and distribution of monitoring points are as follows.

In order to effectively assess the current sewage collection capacity and load in the dry area of the study area, authorities quantitatively evaluate the water equilibrium relationship between the WWTP systems. The dry weather data analysis is based on the monitoring data obtained by the online discharge monitoring equipment (compare Fig. 5.12). A data set is effective only if it is not affected by the rainfall 48 h after the end of the rainfall, and dry weather not less than 7 consecutive days. Through the statistical analysis of the effective data of each monitoring station a number of main parameters are calculated. They cover the daily average flow rate, the maximum water level and the average flow rate are obtained, and the daily average flow rate, peak flow rate and maximum water level in dry weather. The following map of the dry water volume can be obtained based on the analysis of dry weather of the sewage entering the plant in the system in the study area. The equilibrium diagram illustrates the relationship between the water volume of the existing drainage system, and lays a foundation for the water supply in the next step of the joint operating work, and also characterizes the current imbalance of water volume in each area.

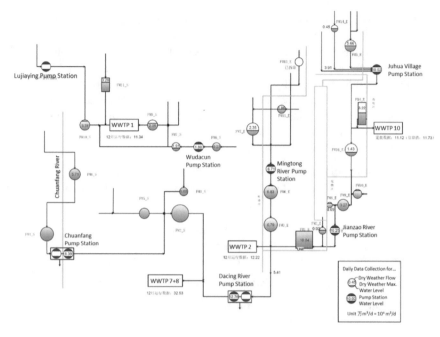

Fig. 5.12 Dry weather analysis of drainage system of WWTPs 1, 2, 7+8 and 10

5.3.3   Model Evaluation and Decision

5.3.3.1   Model Decision of the Storage Project in Western Catchment

With the rapid development of urbanization, the urban drainage system will become more and more complex and larger. The traditional design analysis method is not systematic, the theoretical calculation derivation is different from the actual management operation, and it cannot provide good technical support for the design and reconstruction of urban drainage facilities. Based on the drainage network hydraulic model technology, the evaluation and decision-making system was first applied in the feasibility study of the "Kunming main city western film retention facility project" (outside the second ring). The model was used to analyse the current operation of the area and comprehensively select various design schemes, and scientifically and rationally propose the design and transformation plan of the district.

The area of the main city west area is 66.9 km², of which the area of the diversion system is 33.79 km², and the area of the combined sewer system is 32.65 km², in total 22 combined sewer overflows.

The drainage system of the western catchment could be comprehensively and meticulously analysed by the development of the hydraulic model of the drainage network. The set up and implementation of the model is based on the following:

- basic data collection and analysis of the pipeline network,

- dynamic data monitoring,
- mapping of the foundation structure of the hydraulic model of the drainage network system,
- model parameter mechanism research,
- model checking and verification.

After setting up an operational model, authorities were able to finally applying the hydraulic model to complete the current situation analysis, optimization scheduling and design check.

Firstly, through the establishment of the drainage system hydraulic model, it systematically points out the data problems such as dislocation, reverse slope, large pipe connection, no downstream connection, rain and sewage mixing, and sewage direct discharge into open channels. This improves the accuracy, timeliness and completeness of the drainage facilities database in Kunming. Secondly, the areas with the most serious overflow pollution in the system are determined. Thirdly, by the development of the model project, the resulting data transfer to and application of the project results in the operation management, optimization planning and design transformation of the drainage system in Kunming can be realized.

Based on the GIS data of the existing drainage pipe network of Kunming City, the project establishes a hydraulic model for the west part of the main city, combined with the measured water quantity and water quality data, simulates the hydraulic model of the current drainage system, and evaluates its drainage capacity and pollutant control effect. The feasibility and necessity analysis of storm water and sewage diversion and construction of combined retention facilities are carried out. The optimized engineering design is completed taking into consideration the analysis of environmental and economic benefits. The core requirement is that the total pollutant load discharged into the water body in the drainage area of the combined retention facility is not greater than the discharge from the new diversion or separated system.

The model is used to analyse the pollutant reduction efficiency of the retention facilities in different locations, different water discharge nodes, different design volumes, etc. The comprehensive design ratio is selected to achieve the optimal design of the project. The design scheme and the rainwater and sewage separation within the district are compared with each other to determine the economic and environmental benefits, and finally determine the project implementation plan (compare Fig. 5.13).

Based on the simulation and evaluation, it is proposed to set up three retention facilities of 24,000, 16,000 and 14,000 m^3 respectively to solve the problem of combined sewer overflow pollution in the pumping station areas of Wangjiaqiao and Zhenghe road. In other combined sewer areas, due to the small overflow pollution or the constraints of land use, the new regulation and storage facilities are not economically advised. It is recommended to reduce this part of the combined sewer overflow pollution by optimizing the operation of the western catchment and implementing the urban village reconstruction.

Using the model for long-term simulation calculation, after the implementation of the Western Catchment retention facility project, the annual intercepted combined

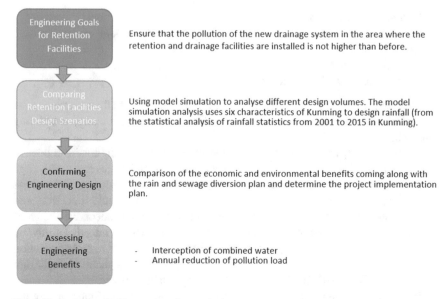

Fig. 5.13 Retention facility engineering analysis

sewage volume is 1.129 million m³, and the total amount of overflow sewage in the combined area is reduced from 32.2% of the total annual runoff in the area to 17.9%. The total amount of combined overflow pollution is 199.6 t COD$_{cr}$/yr, and the engineering benefits are very obvious.

The model-based evaluation and decision-making technology has been applied in the feasibility study of the *Kunming main city south section retention facility project* (outside the second ring) and the *Dianchi basin precision pollution control decision-making project*. The former has carried out a powerful demonstration on the necessity of the construction of the storage facilities in the southern catchment. The latter carried out a simulation analysis of the water quality evolution of the Dianchi Lake, the total water pollution load of the Weishui River, and the total pollution load of the ditch. Based on the water quality response analysis, it guided the overall and orderly construction of the Dianchi Lake treatment project.

5.3.3.2 Accurate Pollution Control Decision in Panlongjiang Catchment

Based on the basic data of river basin key management projects, plots, drainage, and rivers, and monitoring data, the project will establish a land pollution load flux model in Panlongjiang catchment, a three-dimensional water quality-hydrodynamic model of Panlongjiang and a land-water response model of the Panlongjiang catchment (compare Fig. 5.14). These models identify the priority section of the Panlong River and the corresponding water quality contribution, propose effective engineering

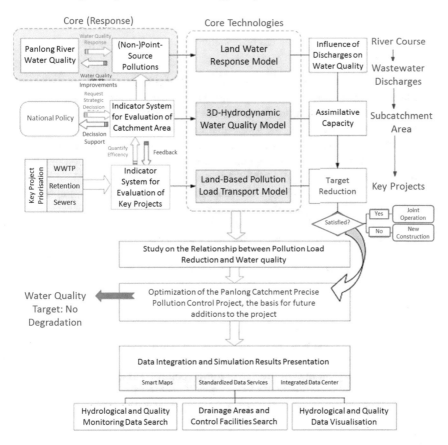

Fig. 5.14 Project technology roadmap at the Panlong River

projects and key treatment areas for "precise pollution control", develop water pollution control engineering assessment and accurate pollution control decision system in Panlongjiang catchment. It realizes the benefit evaluation of key management engineering systems after the implementation of the "13th Five-Year Plan", and form a key management project library that needs to be added or upgraded in the "14th Five-Year Plan" Panlongjiang catchment to provide a basis for scientifically and effectively carrying out water pollution prevention and control in Dianchi Lake Basin.

Two WWTPs, seven water pumping stations, seven retention facilities, and a sewage system of about 689.59 km and a drainage channel length of about 142.93 km have been built in the study area (based on the latest exploration data of the year 2018 in the Panlongjiang catchment). The sewage system mainly includes three types: a separated system for wastewater and rainwater and a combined sewer system. The total wastewater sewage system is 333.97 km, the rainwater system is 350.06 km, and the combined sewage is 3.74 km long.

The technical set up generally follows the sequence of "river course – discharge - sub-zone - key project". The analysis steps for the components are described as follows:

River course: Identify the section of water quality change, establish the response relationship between the discharge and the change of water quality along the river course, and determine the comprehensive index system and quantitative requirements for achieving the water quality goal of the area;

Discharge: Monitor the water flow and load of key discharges, and determine the important discharges affecting specific sections and their contribution rate based on the water quality response with the Panlong River;

Sub-area: Identify the sub-area area importing into the discharges, monitor and simulate the flow and load of the sub-area;

Key projects: Establish the evaluation index system for the project, focus on the impacts of the project on the water flow and load for the sub-area and the discharge, as well as its impacts on the water quality for specific sections of the Panlong River, and propose the quantitative index requirements for optimization or new construction.

For the model evaluation the evaluation indicators of drainage facilities were carefully analysed and calculated. Also, comprehensive assessment of the area and the convergence of water and land were conducted. The results systematically reflect the benefits of engineering governance (compare Fig. 5.15).

Based on the overall idea of precise pollution control and the support of models, it is possible to carry out simulation for comprehensive analysis and evaluation of various management scenarios. Scenarios are including

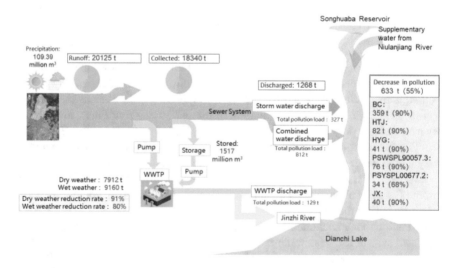

Fig. 5.15 Results of the comprehensive 2018 assessment

- the overflow analysis of main discharges under different rainfall conditions,
- the assessment of primary strengthened pollution reduction and the water quality improvement of the 5th WWTP,
- the water quality impact assessment of the Panlong River because of the discharge from the tail water of the 4th and 5th WWTP, and
- the water quality impact assessment of the Panlong River under the operation of the 14th WWTP and so on.

The scenario modelling can support the scientific decision-making process for relevant treatment plans. Take the water quality impact assessment of the Panlong River under the operation of the 14th WWTP as an example.

According to the project feasibility study design document of the 14th WWTP, the completion of the 14th WWTP may largely influence the wastewater treatment and the water quality in the Panlong River area. The implementation of the measures for the 14th WWTP is divided into 2 stages: near future and long-term. So here we show two important scenarios based on the relevant construction documents, namely:

(1) *The near future operation:*
    On the basis of the current situation in 2017, the overloaded wastewater of the 5th WWTP will be transferred to the 14th WWTP. Under the recent conditions of 14th WWTP, its advanced treatment capacity is 100,000 m^3 per day (The tail water does not enter the Panlong River), and its primary treatment capacity after reconstruction is 400,000 m^3 per day, with water entering the Panlong River. The advanced treatment capacity of the 5th WWTP has dropped from 240,000 m^3 per day to the design capacity as 185,000 m^3 per day (the water does not enter Panlong River just like the current situation in 2017). The situation of the 4th WWTP and other drainage facilities remain unchanged.

(2) Based on the near future operation of the 14th WWTP, we improve the capacity of advanced treatment of the 14th WWTP from 100,000 to 200,000 m^3 per day, and the tail water still does not enter the Panlong River. At the same time, closing the 4th WWTP and transferring its wastewater to the 5th WWTP for treatment. By setting the above conditions to the model and performing simulation evaluation, comprehensive evaluation and comparison of engineering benefits can be carried out, and the improvement effect on water quality can be objectively evaluated (compare Fig. 5.16).

5.3.4 System Joint Operation

With the improvement of drainage facilities in Kunming City, four relatively independent drainage areas have been delimited in the main city, including the drainage systems of the western, northern, southern and south-eastern area. During the 12th Five-Year Plan period, the systematic joint operation technologic research and engineering demonstration were first conducted in the drainage system of the western

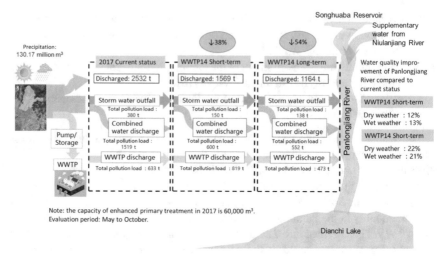

Fig. 5.16 Evaluation of improvement effect on water quality

area, which focused on the improvement of water environment and water ecology of the old canal for grain transportation.

At the beginning of the project, there was only the 3rd WWTP, located in the lower reaches of the old canal and adjacent to Caohai. Four pumping stations, Tudui, Zhangfeng, Wulong River and Zhuangfang were set outside and could be directly connected to the fine grille gallery of the 3rd WWTP via a pressure pipe. With the development of the city, the 9th WWTP was built in the northwest area in 2014 in order to alleviate the load of the 3rd WWTP. Besides, some dividing weirs were set up on the three main trunk sewer that originally led to the 3rd WWTP, so that the wastewater in the northwest area was preferentially sent to the 9th WWTP, and the excess wastewater could be diverted to the 3rd WWTP. In the dry season, the wastewater in the area can be fully collected and treated. Due to the difficulty in implementing the transformation to separate system, the combined sewer system still exists. There are five main overflow structures for combined sewer system along the old canal. During the 12th Five-Year Plan period, in order to alleviate the problem of rainwater overflows over the dam in the rainy season, five rainwater retention facilities were installed at the end of the main sewer system, with a total retention capacity of about 56,000 m³. A complex drainage system was formed in the study area, called Plant-Station-Pool-Net. Flows in the service area of the 9th WWTP can be described clearly. There are two independent treatment processes in the service area of the 3rd WWTP as well as several pumping station and retention facilities. Thus, flows analysis in this catchment is more complex.

Firstly, based on the results of field survey and drainage model, 19 online liquid level monitoring devices were installed at the end overflow, diversion, confluence, low-lying and sensitive control points, and the data can be transmitted to the remote operating centre in real time via wireless. It is convenient to monitor the load in real-time and control the drainage system in real time. This can provide support for the analysis and control for the plant-network joint operating model.

Secondly, through the transformation of a self-control system and the construction of a data acquisition system, the function of data acquisition and remote monitoring of the two WWTPs, five retention facilities and four pump stations in the West Zone was realized.

Thirdly, based on the analysis of the drainage GIS data and the historical data of the monitoring points, the basic framework of the drainage facilities operating control in the western area was built. The control strategy was determined through the analysis of operating parameters, the load of WWTP, the operation load of sewer system, and the overflow of the key outlets. Around the goal of optimized load distribution and discharge overflow control, we developed a three hierarchical control strategy and algorithm, consisting of:

- plant-plant load distribution control,
- plant-station flow balance control,
- station-pool-network coordinated control.

Among them, the plant-plant load distribution is controlled using the water level at the fixed weir. This control is aiming to solve the problem of balancing the load between the plant and plant. The plant-station flow balance control focuses on the priority of the peripheral pumping station along the WWTP (compare Fig. 5.17). The plant-pool-network coordinated control is based on the online water level of the

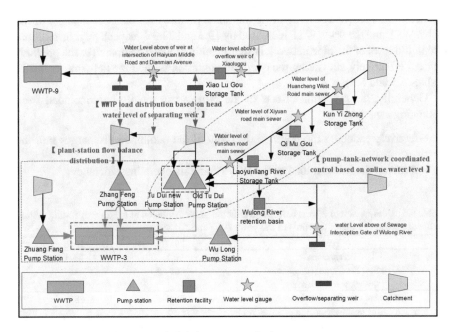

Fig. 5.17 Diagram of joint control of drainage system in the western area

pipe network and is calculating when the water enters the retention facility and when it is emptied.

After the basic preparatory work, the selection of the core operating strategy and the development of algorithms, the hardware integration was carried out in addition to the digitization of the drainage system. The construction management department had organized the basis for the system modelling. With the construction of the software platform, the comprehensive management and analysis of the drainage GIS data can be realized. The process elements of the drainage facilities in the western area can be fully monitored. They include pump stations, WWTP, retention facilities, sewage system monitoring points and so on. The operating instructions can be automatically generated and issued, and the historical operating records are logged and can be checked.

The pilot project started construction in May 2017 and was completed on August 1, 2017. After nearly one month of commissioning, it entered normal operation on September 1. A comparative analysis of the operational effects of dry seasons during the same period in 2017 and 2018 was carried out. In the case of similar rainfall, the total overflow of the main discharges in this area decreased from 640,000 tons to 350,000 tons, with the reduction rate reached 45.3%. During this same period, the treatment capacity of the 3rd WWTP and the 9th WWTP increased by 7.1% and 34.2% respectively. Due to the lack of complete rainy season operational data for the demonstration project, a statistical analysis for three months around August 1, 2017 of the operational effects was carried out. In the case of similar rainfall, the total overflow of the main discharges decreased from 5.98 to 5.65 million tons, with the reduction rate reached 5.5%. During this same period, the treatment capacity of the 3rd WWTP and the 9th WWTP increased by 19.3 and 22.9% respectively. In addition, the liquid level of the pipe network is significantly reduced compared to the previous period, effectively alleviating the overload condition of the pipe network (compare Tables 5.5 and 5.6).

Generally, the operation effects of the project can mainly be concluded from following aspects:

- effectively reducing the polluted overflow of the combined discharges into the river;
- improving the influent load of the WWTP and its operating efficiency;

Table 5.5 Comparison of effects before and after the implementation of the joint operating system in the dry season (January to April) of the years 2017 and 2018

Year	2017	2018
Precipitation	90,1	86,4
Overflow Discharge (10,000 tons)	64	35
Overflow reduction rate	45,3	
Treatment amount of the 3rd WWTP	2684	2875
Increase rate of treatment amount of 3rd WWTP	7,1	
Treatment amount of the 9th WWTP	628	843
Increase rate of treatment amount of 9th WWTP	34,2	

Table 5.6 Comparison of effects before and after the implementation of the joint operating system in the rainy season of 2017

Date	May to July		August to October
Rainfall	474,3		477,6
Overflow discharge (10,000 tons)	598		565
Overflow reduction rate		5,5	
Treatment amount of the 3rd WWTP	1980		2362
Increase rate of treatment amount of 3rd WWTP		19,3	
Treatment amount of the 9th WWTP	572		703
Increase rate of treatment amount of 9th WWTP		22,9	

- reducing the overload rate of the drainage pipe network, and therefore improving the flood prevention and emergency response capability in the rainy season;
- sharing information and providing feed-forward support for process optimization of the water purification plant and emergency response in the rainy season, which helps to ensure stable operation of the WWTP.

The pilot research and engineering practice in the western area develops a multi-facility joint operating control algorithm and systematic integration scheme with distinguished characteristics of real-time, robustness and universality. It provides underlying technical support to the construction of integrated efficient system in other drainage areas in Kunming and even to the operating control to the water circulation system of the Dianchi Lake.

5.3.5   Summary of Application Benefits

The digitization of the management of the urban drainage system in Kunming, as described above, help to build an integrated management of plant-network-river-lake system. The application benefits are mainly the following:

(1)   Accurate and comprehensive reflection of the current status of Dianchi treatment facilities

This component can promote periodic and iteratively updated on-site data verification and supplementary surveying based on the existing drainage network data of Kunming City and the "Technical Specifications for Data Management of Urban Drainage Facilities in Yunnan Province". It can also realize data standardization and consolidation, as well as carrying out topology inspection and correction, and finally building a GIS database of drainage facilities. The GIS basically reflects the status quo. It helps to operate an effective online monitoring network to accurately reflect the operation dynamics of drainage facilities, which can provide reliable information to management and decision-making process. In the application process, establishing a data sharing and supplementary update mechanism can further improve the data quality and use value.

(2)   Systematic diagnosis and evaluation of construction and renovation plan for Dianchi

Based on the mathematical model, this element can establish a specialized analysis and evaluation. It helps to carry out simulation and quantitative evaluation of the engineering schemes for Dianchi Lake governance. This system can systematically analyse and scientifically diagnose the effects of the facility construction including:

- analyse the problems and defects of the existing renovation plan, and diagnose the risk of relevant decisions;
- provide decision making support for planning and transformation project, large-scale project, operation scheduling, linkage control etc.

The assistant management staff can quantitatively evaluate the load and operating status of the water supply and drainage system with the joint application of online monitoring and model, and then evaluating the relevant planning and transformation plan, improving the application benefit.

(3)   Improvement of the operating efficiency of large-scale projects and optimize investment

Based on the simulation optimization analysis method of complex system, a new planning and design evaluation and optimization process is established. The problem diagnosis, scheme formulation and engineering decision-making process of the drainage system are assisted by means of monitoring and modelling with the purpose of improving the water quality of Dianchi Lake and the overall operating efficiency of the drainage system. Meanwhile, it carries out the research on joint control. It also improves the reliability and effectiveness of the overall operation of the drainage system. The system helps with the overall optimization and synergy scheme by a multi-objective optimization method. It also fully mines the potential of existing drainage facilities, sewage systems and WWTP and their existing capacities to reduce unnecessary or inefficient construction investment.

(4)   Improve the decision-making level of the planning and construction of pollution control projects in Dianchi Lake Basin

Whether it is possible to improve the planning and design decision-making level of major projects depends largely on whether the related status data of the drainage facilities are truly and completely grasped. The improvement also depends on whether the overall operational data are truly diagnosed and evaluated during the engineering planning and construction decision period. This program has designed a new and improved database, necessary hardware support, scientific decision-making mode, and visualized control system. It helps the decision-makers of the investment company to grasp the whole process information of the related project as detailed and comprehensive as possible. The process information is a solid foundation for big data analysis. Based on a large number of updated data and advanced model analysis methods, it is possible now to comprehensively evaluate the operating situation of system, optimize relevant details, thereby improving the decision-making level of engineering planning and construction.

(5) Build a long-term dynamic decision-making model combining technology and management

The complexity of the drainage system determine that the state of the drainage system is dynamic and stays to some extent uncertain. This program builds a long-term dynamic decision-making model. On the one hand, it helps to establish an iterative data update mechanism and an online monitoring and early warning system for drainage facilities to form a drainage monitoring network. It can truly assess the existing status of the drainage facility and operates the monitoring and early warning. On the other hand, it establishes a decision support model with joint application of multiple models. It quantitatively simulates and analyses the engineering planning scheme and gives decision-making suggestions. At the same time, it can use online monitoring, model simulation, multi-objective optimization, big data analysis and other means to forecast the status of drainage facilities in the future, comprehensively and systematically analyse the historic, current and future development of the Dianchi pollution control system to promote the scientific decision-making.

5.4 Conclusion and Prospects

Integrating the monitoring and model system to form an intelligent decision-making and control system that can quantitatively evaluate and support decision making, and to realize the unified integration of the whole process information of "source-network-station-WWTP-river-lake" system can effectively support the refined, scientific, systematic, intensive governance of Dianchi and serve the long-term water quality improvement. In the process of Dianchi Lake management, it is of great significance to employ advanced information technology and the core technology system including basic data, online monitoring, simulation, evaluation and decision-making and intelligent control, and build an integrated management and control mechanism of source-network-station-WWTP-river-lake system to effectively support engineering decision-making and operation control.

Especially with the further growth of urban areas of Kunming, the drainage system is becoming more and more complex and larger, which increases the pressure of planning and decision-making. The existing local problem analysis and diagnosis methods have been unable to meet the needs of optimization and improvement for complex drainage system. In order to increase effectiveness of the large number of existing drainage facilities and engineering projects it is necessary to implement new technology. They aim at a number of tasks such as

- objectively evaluate the operational efficiency of the existing systems,
- propose necessary new construction and reconstruction projects, and
- provide clear, comprehensive and practical instructions for the future drainage information construction in Kunming.

To do so, we comprehensively use advanced technologies such as GIS technology, network communication, industrial automatic control and drainage system simulation. These are the core components to build a comprehensive management and control platform capable of managing a large amount of spatial and attribute data of the drainage system in a long-term, effective and dynamic manner. On the basis of the implemented functions and acquired data of the new information system, we need to form a comprehensive database of drainage system in the downtown area. We also construct various business modules and professional analysis modules required by the digital management of the drainage system. With them, we step by step study and explore the intelligent control mode of the drainage network. Finally, we can achieve the construction goal of building the "source-network-station-WWTP-river-lake integrated management system". This will help to comprehensively improve the diagnostic analysis level, planning decision-making level and performance evaluation ability of the drainage facilities in Kunming.

In the planning management and engineering decision-making process of the drainage system in Kunming, we should continuously explore the planning and control mode of "source-network-station-WWTP-river-lake" integration to realize the goal of overall decision diagnosis and optimization. We should promote continuous improvement of digitization to motivate the transformation of management and control and realize scientific decision-making and intelligent operation of drainage facilities. We do that by providing a unified decision-making analysis platform for Dianchi water environmental governance, realizing the whole process information management of "source-network-station-WWTP-river-lake" system. We should also provide reliable, unified and detailed data interface and systematic system for the call of relevant business systems, and finally promote the transformation and improvement of urban management mode.

In the iterative improvement process of the integrated management system, it is also necessary to actively draw on international advanced experience, especially the close cooperation between SINOWATER and FiW in Sino-German cooperation projects. We should further promote the development of the drainage system in the future in Kunming. Including but not limited to the following points:

(1) The acquisition of data is the basis and core of all work, its collection and transmission need to be continuously implemented in the project construction. And it is necessary to ensure capital investment.

(2) Applying models to guide designing process and performing systematic and scientific decision-making are essential means.

(3) Continuously and steadily promote rain and sewage system separation, LID source reduction measures of rainwater runoff combined with the long-term urban expansion and transformation.

(4) Improving the efficiency of the sewer system and reduce the difficulty of operation and maintenance. Separation of wastewater from rivers by setting up facilities within the open sewage channels

(5) Sewage system inspection, infiltration protection, operation and maintenance of sewage systems are very important.

(6) The implementation of new treating methods like soil/wetland treatment should be followed by the match of adjusted retention facilities and the scale of WWTP.

(7) The initial rainwater interception and limit of the separated system should be sent to the WWTP.

(8) One of the objectives of the next stage of the WTTP is the treatment of micro-pollutants.

Integrating the monitoring and model system to form an intelligent decision-making and control system that can quantitatively evaluate and support decision making, and to realize the unified integration of the whole process

To conclude, the establishment of a scientific and effective "source-network-station-WWTP-river-lake" integrated management system and its rational application to actual planning and decision-making is not a one-step process. It requires the long-term and unremitting efforts of decision makers, planning designers, professional technical teams and operating management staff to build an integrated drainage management system, which is scientific, practical and can meet the local technology needs.

References

1. The Ministry of Housing and Urban-Rural Development of the People's Republic of China(MOHURD). (2015) City Office Letter No 635.
2. The Technical Standard for Drainage Data Management of Yunnan Province (DBJ53/T-93-2018). Yunnan: The Ministry of Housing and Urban-Rural Development of Yunnan Province, 2018.
3. The Technical Standard for Online Data Collection of Urban Drainage Facilities in Yunnan Provice (DBJ53/T-94-2018). Yunnan: The Ministry of Housing and Urban-Rural Development of Yunnan Province, 2018.

Chapter 6
Treatment of Wastewater with High Proportion of Industrial Wastewater with Extreme Requirements on the Elimination of Nutrients

Max Dohmann, Yunbo Yun, and Vivien Lee

6.1 Introduction

China's industrial and economic development over the past 20 years has particularly affected the provinces on the east coast of China. The absolute focus of development has been the Greater Shanghai region and the Yangtze delta area. This development was accompanied by a high demand for drinking water and industrial water and a correspondingly high volume of wastewater. The inadequate treatment of wastewater from municipalities, industry and agriculture led to a significant deterioration in the quality of water resources. This applies in particular to Lake Tai, the most important regional water resource in the greater Shanghai area. In the summer of 2007, Lake Tai developed very strong eutrophication for the first time, so that the water could no longer be used for drinking water supply for a time. Due to the massive expansion of wastewater treatment with nutrient elimination that has taken place in the meantime, the conditions have improved somewhat.

With the 13th Five-Year Plan of the Chinese government, water protection has gained special political significance since 2016. This included much stricter targets for the water quality of sensitive waters such as Lake Tai. As a result, Jiangsu Province, to which a large part of the lake catchment area belongs, issued new requirements for wastewater treatment. These are specified in the "Discharge Standard of Main Wastewater Pollutants for Municipal Wastewater Treatment Plant & Key Industries of Taihu Lake Area". This also applies to the nutrient elimination requirements for the treatment plants in the lake catchment area. Previously, two categories of wastewater treatment plants were considered: Category I wastewater treatment plants with less than 50% industrial wastewater and Category II wastewater treatment plants with more than 50% industrial wastewater. Since May 2018,

M. Dohmann (✉) · Y. Yun · V. Lee
Research Institute for Water and Waste Management, RWTH Aachen University (FiW) e. V., Aachen, Germany
e-mail: dohmann@fiw.rwth-aachen.de

© The Author(s) 2022
M. Dohmann et al. (eds.), *Chinese Water Systems*, Terrestrial Environmental Sciences,
https://doi.org/10.1007/978-3-030-80234-9_6

new requirements have been in effect, which were published in DB32/1072-2018. These apply to all wastewater treatment plants regardless of the respective industrial wastewater content. Limit values for the nutrient parameters nitrogen and phosphorus play a special role. A comparison of the limit values of the guidelines for the Tai Lake catchment area and the original quality specification according to the Chinese Water Protection Directive GB18918-2002 (I-A) was made in Table 6.1. It can be seen that above all the stricter requirements for nitrogen elimination, which have been tightened since 2018, represent a particular challenge for wastewater treatment plants.

A comparison of the minimum requirements applicable in Germany for the effluent quality of municipal wastewater treatment plants of different size classes and Chinese requirements is shown in Table 6.2. It should be noted, however, that

Table 6.1 Requirements for wastewater treatment plant discharges in the Tai Lake area according to Guidelines DB32/1072-2007 and DB32/1072/2018 and process standard I-A according to Chinese Water Protection Directive GB18918-2002

Parameter		DB32/1072-2007			DB32/1072-2018		GB18918-2002
		until 12/31/2007		since 01/01/2008	since 06/01/2018	from the 01/01/2021	SK I-A
		I	II				
COD	[mg/l]	50	60	50	50	40	50
NH₄-N	[mg/l]	5 (8)*	5 (8)*	5 (8)*	4 (6)*	3 (5)*	5(8)*
TN	[mg/l]	20	15	15	12 (15)*	10 (12)*	15
TP	[mg/l]	0,5	0,5	0,5	0,5	0,3	0,5
* for temperatures ≤12°C							

Table 6.2 Comparison of German minimum requirements for wastewater treatment plant effluents according to the Waste Water Ordinance (ABWV 2004) and Chinese requirements (see also Table 6.1)

Parameter		Size class					SK I-A	DB32-2018
		1	2	3	4	5		
Größen-klasse	(kg/d BOD)	<60	60-300	300-600	600-6.000	>6000	-	-
COD	[mg/l]	150	110	90	90	75	50	50
BOD	[mg/l]	40	25	20	20	15	10	-
NH₄-N	[mg/l]	-	-	10	10	10	5	4
TN	[mg/l]	-	-	-	18	13	15	12
TP	[mg/l]	-	-	-	2	1	0,5	0,5

the German minimum requirements in Table 6.2 are based on an emissions analysis only. Higher requirements may result in Germany from emission aspects. Nevertheless, the comparison in Table 6.2 shows the particularly high Chinese requirements for nitrogen elimination in the Tai Lake area that have been in force since 2018.

The background to the investigations described in this report was provided at the beginning of 2016 by contacts between German researchers within the framework of the Chinese Major Water Programme and the Chinese company Huayan Water Affairs Co., Ltd., which is responsible for water supply and wastewater disposal in the Suzhou Industrial Park on Lake Tai. Discussions held with the regional water authorities during this period revealed that in future, extreme demands will be placed above all on nitrogen elimination at the Tai Lake treatment plants. The talk was of a discharge value for total nitrogen of less than 6 mg/L. The discussed effluent value for the COD should be 30 mg/L. This resulted in a German-Chinese research project with the aim of investigating technical possibilities to meet such extreme requirements. The research project was carried out with the participation of the Research Institute for Water and Waste Management at the RWTH Aachen University (FiW) and the companies Atemis GmbH, Aachen and EvU GmbH, Gröditz on the German side and the University of Shanghai for Science and Technology and the company Huayan Water Affairs Co. The German Ministry of Education and Research (BMBF) provided funding for the German research work carried out in this project as part of the German-Chinese Major Water Cooperation.

6.2 Industrial Wastewater Treatment in China

For the year 2015, the share of industrial wastewater in China was stated as 26.2% of total wastewater (Xiang 2017). Most industrial wastewater in China is discharged into municipal sewerage systems or sewage treatment plants. However, it can be assumed that the actual share of industrial wastewater is higher. Not all small and medium-sized industrial enterprises are included in the statistics. In 2015, industrial wastewater in China had a COD content of 25.8% and a NH4-N content of 13.9%, compared to 74.9 and 86.1% of domestic wastewater respectively (Xiang 2017). However, even a certain increase in the industrial wastewater fraction would not adequately reflect the major environmental impacts caused compared to domestic wastewater. Many industrial effluents are contaminated with pollutants that are not adequately retained by treatment in sewage treatment plants and thus lead to disproportionate water pollution.

The following are the characteristics of industrial wastewater compared to domestic wastewater:

– one-sided composition of the wastewater constituents
– high concentration of individual wastewater constituents
– high fluctuations in the concentrations
– intermittent wastewater flow and pH fluctuations

– elevated temperatures
– dyes or persistent organic substances.

A particular problem is the treatment of industrial wastewater with high organic loads that are not or only with difficulty biodegradable and the load of heavy metals. Up to now, compliance with specified concentrations in the effluent of many Chinese wastewater treatment plants has only been possible by diluting industrial wastewater with less or non-contaminated wastewater. The consequence must be that in the future highly contaminated partial streams of industrial wastewater must be specifically pre-treated before further treatment.

There are various possibilities for the joint or separate treatment of municipal and industrial wastewater. Figure 6.1 shows the four basic concepts for the arrangement of the treatment plants. All four concepts have been implemented in German wastewater practice. Concept A is certainly only suitable for very large industrial plants. This also applies to concept D, where municipal wastewater can be used for improved treatment of industrial wastewater. In China, most of the industrial wastewater is co-treated in municipal wastewater treatment plants.

When industrial wastewater is discharged into a municipal sewage system, it must be ensured that the requirements for the effluent quality of the plants are met. Beyond the aspects of water protection, it must be ensured that the sewage system is not damaged by industrial wastewater. Structural damage can be expected, for example, if high sulphate or ammonium concentrations are present. In order to protect public sewage systems from such damage, appropriate concentration limits should therefore be observed.

For larger industrial plants with wastewater from different production processes, a material flow separation with correspondingly differentiated wastewater treatment has proven to be effective in Germany for water protection reasons. In this way, improved recycling of wastewater and wastewater constituents can be achieved. The amount of recycled and reused wastewater has so far remained low in China, with only

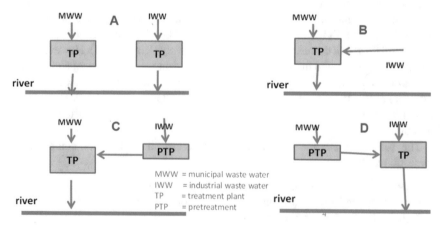

Fig. 6.1 Basic concepts for the treatment of municipal and industrial wastewater (Dohmann 2014)

10% of treated industrial wastewater (Xu 2016). In recent years, several guidelines for water recycling have been published in China (Xu 2016). For example, the Water Pollution Prevention Action Plan and Control Action Plan of April 2015 states that by 2020 the rate of use of recycled water should be 20% in water-poor cities and 30% in the Jing-Jin-Ji area, which is a particular challenge for planning new industrial wastewater systems in China and complementing existing ones.

6.3   The Study Area and Investigated Wastewater Treatment Plant

The study area, the Suzhou Industrial Park (SIP), covers part of the urban area of Suzhou to the west of Shanghai (see Fig. 6.2). In terms of water management, the region is characterized by Lake Tai, which serves as an important water resource for the metropolitan areas of Suzhou, Wuxi, Changzhou and Nanjing. The city of Suzhou is particularly well known for its numerous canals, which is why it was given the name "Venice of China". In recent decades, the economic output in Suzhou has grown very strongly with the settlement of many industrial companies. The Suzhou Industrial Park was founded in 1992 and covers a catchment area of 278 km^2 with over 2,000 industrial companies.

Suzhou Industrial Park has two wastewater treatment plants of roughly the same size, which treat the industrial wastewater and domestic wastewater of the approximately 800,000 inhabitants living in the park's catchment area. The investigations carried out in the research project related to the Suzhou II WWTP, located in the southern part of the park. This plant, commissioned in 2010, was designed for a daily wastewater inflow of 150,000 m^3. In 2017, the average daily inflow was about 172,000 m^3. The share of treated industrial waste water accounted for almost 60% of

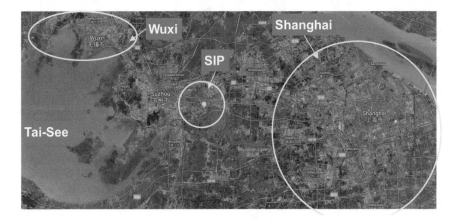

Fig. 6.2 Regional location of Suzhou Industrial Park (Google Maps 2018)

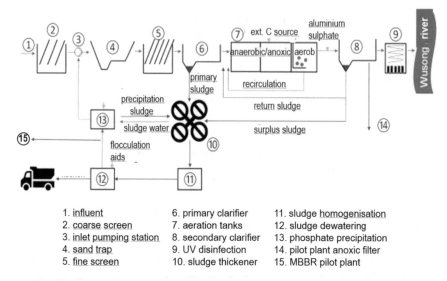

1. influent
2. coarse screen
3. inlet pumping station
4. sand trap
5. fine screen

6. primary clarifier
7. aeration tanks
8. secondary clarifier
9. UV disinfection
10. sludge thickener

11. sludge homogenisation
12. sludge dewatering
13. phosphate precipitation
14. pilot plant anoxic filter
15. MBBR pilot plant

Fig. 6.3 Process diagram of the investigated Suzhou II wastewater treatment plant in the Suzhou Industrial Park (Lee 2018)

the total inflow. The basic concept of this wastewater treatment plant corresponds to the concept in Fig. 6.1C, although, unlike in Fig. 6.1, all sewers and the wastewater treatment plant are operated by the private company Huayan Water.

Figure 6.3 shows the process diagram of Suzhou II WWTP and the location of the two pilot plants, which are reported on in more detail in Chap. 6.3.2 and 6.3.3.

The treatment plant has a classical process engineering for municipal wastewater. In contrast to most Chinese wastewater treatment plants, a sedimentation stage takes place before the biological treatment according to the AAO system. The wastewater treatment plant is equipped with extensive online measuring technology. First of all, pH value, temperature, COD, NH4-N, TN and TP are measured behind the fine screen. In the aerobic part of the biological stage, the online values of O2, redox potential, solids content, TN and TP are measured, while at the UV disinfection stage pH, temperature, COD, filterable substances, NH4-N, TN and TP are recorded. The phosphorus elimination is carried out biologically and by means of a simultaneous precipitation with aluminium sulphate.

The sludge produced in the sedimentation stage is thickened together with the excess sludge of the biological stage and dewatered mechanically with centrifuges with the addition of flocculation aids. The sludge is then dried in an external drying plant. The sludge dried to 90% dry substance is then incinerated in a thermal power plant. The process water from the thickeners and the mechanical sludge dewatering is treated in a precipitation plant before being returned to the inlet pumping station for phosphate elimination.

Table 6.3 shows the design values for the treatment plant and the mean inflow values in 2017 and in the investigation phase. During the investigations, the pollutant

Table 6.3 Inflow values of the investigated Suzhou II wastewater treatment plant (Lee 2018)

Parameter		planned loads	inflow values 2017	average inflow values in the project phase
average inflow rate	[m³/d]	150.000	172.092	163.322
dry weather inflow	[m³/d]	130.000	-	-
maximum inflow	[m³/h]	8125	-	-
BOD	[mg/l]	20-250	232	288
COD	[mg/l]	500	447	571
TN	[mg/l]	15-55	37	44
NO₃-N	[mg/l]		3	2
TP	[mg/l]	0,25-6	5	7
filterable SS	[mg/l]	350	194	202

Table 6.4 Requirements for industrial wastewater when discharged into the sewerage system in Suzhou Industrial Park according to CJ343-2010

COD [mg/l]	BOD [mg/l]	filter.SS [mg/l]	pH	NH₄-N [mg/l]	TN [mg/l]	TP [mg/l]	Oil [mg/l]
≤500	≤350	≤400	6,7~9,5	≤45	≤70	≤8	≤20

load of the plant corresponded to an estimated 665.000 population equivalents. Due to the further expansion of industry and the planned further expansion of residential development in the industrial park, the Suzhou II wastewater treatment plant is to be expanded to a daily inflow load of 300.000 m³.

For the industrial wastewater generated in the Suzhou Industrial Park, requirements must be met before it is discharged into the sewerage system. According to the Chinese guideline CJ343-2010, these relate to the pH value, organic pollutants, nutrients and oil content. The corresponding limit values are shown in Table 6.4.

6.4   Investigation Approaches

The overall objective of the research project was to reduce the nutrient content of the effluent from the Suzhou II sewage treatment plant. The main focus was the elimination of nitrogen with a target concentration of 10 or even 6 mg/L. In connection with the elimination of the organic effluent values, which also had to be improved, the investigations in Germany were to make use of the technological experience gained and German administrative management approaches.

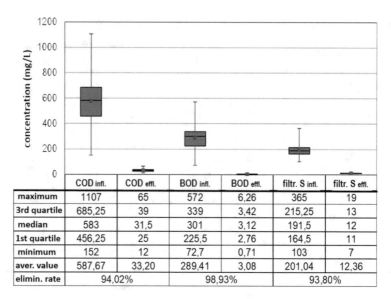

	COD infl.	COD effl.	BOD infl.	BOD effl.	filtr. S infl.	filtr. S effl.
maximum	1107	65	572	6,26	365	19
3rd quartile	685,25	39	339	3,42	215,25	13
median	583	31,5	301	3,12	191,5	12
1st quartile	456,25	25	225,5	2,76	164,5	11
minimum	152	12	72,7	0,71	103	7
aver. value	587,67	33,20	289,41	3,08	201,04	12,36
elimin. rate	94,02%		98,93%		93,80%	

Fig. 6.4 Inflow and outflow values for the parameters COD, BOD and filtrateable substances for the Suzhou II wastewater treatment plant in the 2017/2018 investigation period (Lee 2018)

The necessity of further treatment effects for the Suzhou II wastewater treatment plant can be demonstrated by the previous inflow and outflow data of the plant. Figures 6.4 and 6.5 show the concentration ranges and elimination rates for the parameters COD, BOD5, filterable substances and the nutrient parameters nitrogen and phosphorus.

Figure 6.4 shows that although the mean COD effluent value of 31.5 mg/L was well below the requirements of >50 mg/L applicable since 2008, the maximum value measured was 65 mg/L. If the future treatment capacities of the wastewater treatment plant are to ensure a maximum COD discharge value of 30 mg/L, there is a need for a significantly improved COD elimination. Such a value of 30 mg/L could not be achieved by a solids-free effluent alone, but requires above all process engineering measures in the biological treatment stage.

Figure 6.5 illustrates the treatment performance of the Suzhou II WWTP for the nutrients nitrogen and phosphorus. In the case of phosphorus elimination, compliance with the requirement of 0.5 mg/L total phosphorus in the effluent, which is valid until 2021 (see Table 6.1), does not seem to pose a difficult problem. Even the effluent value <0.3 mg/L applicable after 2021 appears to be well possible with an optimization of the previous elimination mechanisms. The maximum measured value of 1.03 mg/L mentioned in Fig. 6.5 should certainly be considered as the outlier value.

In the case of nitrogen elimination, on the other hand, the situation is different. In the case of a massive disturbance of nitrification caused by industrial wastewater, there was a short-term extremely high ammonium nitrogen concentration in the wastewater treatment plant effluent during the period under study. This is not

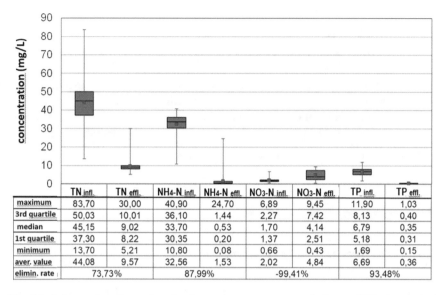

	TN infl.	TN effl.	NH4-N infl.	NH4-N effl.	NO3-N infl.	NO3-N effl.	TP infl.	TP effl.
maximum	83,70	30,00	40,90	24,70	6,89	9,45	11,90	1,03
3rd quartile	50,03	10,01	36,10	1,44	2,27	7,42	8,13	0,40
median	45,15	9,02	33,70	0,53	1,70	4,14	6,79	0,35
1st quartile	37,30	8,22	30,35	0,20	1,37	2,51	5,18	0,31
minimum	13,70	5,21	10,80	0,08	0,66	0,43	1,69	0,15
aver. value	44,08	9,57	32,56	1,53	2,02	4,84	6,69	0,36
elimin. rate	73,73%		87,99%		-99,41%		93,48%	

Fig. 6.5 Inflow and outflow values for the nutrient parameters total nitrogen(TN), Ammonium (NH4-N), nitrate (NO3-N), and total phosphorus (TP) for the Suzhou II wastewater treatment plant in the investigation period 2017/18 (Lee 2018)

unusual for industrial wastewater discharges, but given the size of this treatment plant, this represents a very significant water pollution, which should be avoided in the future by increased monitoring of indirect discharges. The total nitrogen value of the wastewater treatment plant effluent during the period of the above-mentioned disturbance was twice the concentration permitted until 2018.

As the mean value for total nitrogen was at the level of the requirement of 10 mg/L applicable from 2021 onwards according to Table 6.1, it is evident that about half of the measured values recorded were higher and that therefore a significantly improved nitrogen elimination is necessary in the future. This means a significant increase in the denitrification performance of the treatment plant. The discharge value for total nitrogen of 6 mg/L at most, which is the long-term objective, will probably be very difficult to meet in the long term at the nitrate levels shown in Fig. 6.5.

Based on the findings on the treatment performance of the Suzhou II WWTP and the requirements to be met in the future, the following work packages or investigation approaches were carried out within the framework of the research project:

a) Indirect discharger monitoring and management approaches with application of German experience for the recording and analysis of industrial wastewater discharges in the Suzhou Industrial Park

b) Investigations for the process engineering optimisation or completion of the existing Suzhou wastewater treatment plant

   – optimisation of the operation mode of the primary clarification by laboratory tests to control the sedimentation,

- C-dosage to increase denitrification in the biological stage by using German experience and laboratory experiments to test different substrates,
- Nitrogen elimination by downstream anoxic filtration - design, construction and operation of a pilot plant "denitrification filter",
- Nitrogen elimination from the partial flow of sludge water by conception, construction and operation of a SBR-MBBR test plant,
- Nitrogen elimination from the partial flow of sludge water through the design, construction and operation of a test plant for the deammonification.

The investigations were carried out in 2017 and 2018.

6.5   Results of Investigation

6.5.1   Indirect Discharger Monitoring

In Germany, the individual German provinces have issued regulations for the discharge of wastewater into public wastewater systems. The legal background for these so-called indirect discharger regulations is water protection. These ordinances refer to the German Wastewater Ordinance, which covers 57 source areas for wastewater (AbwV 2004). In addition to domestic wastewater, this concerns commercial and industrial wastewater from 56 different sectors. The indirect discharger regulations contain concentrations of various substances contained in the wastewater which must not be exceeded when it is discharged into a public sewerage system or a wastewater treatment plant. When determining these limit concentrations, it was taken into account that these constituents can be at least partially eliminated in wastewater treatment. The monitoring of industrial indirect dischargers in Germany is carried out by the local water authorities.

The German provinces oblige the operators of public wastewater treatment plants to keep a list of industrial enterprises whose waste water is expected to have a significant impact on the wastewater treatment plants or on water bodies. This register is known as the indirect discharger register and must be updated regularly. The establishments are obliged to provide the necessary information. The list shall be submitted to the competent water authority on request. There is decades of experience in Germany in the preparation of indirect discharger registers. The practical basis for this is a special set of rules (DWA 2013).

In addition to the requirements regulated by the water authorities for the discharge of industrial wastewater into public wastewater systems, requirements must also be set for the wastewater to be discharged in the interest of the operator. These requirements are intended to ensure that these discharges do not cause structural damage to the public wastewater systems and to guarantee the operational safety of the plants and the protection of the operating personnel. The corresponding requirements for the quality of wastewater are part of the respective municipal wastewater regulations. The operator of the public wastewater treatment plants is responsible for monitoring the industrial indirect dischargers with regard to these requirements. In Germany, therefore, parallel monitoring measures in the indirect discharge area are possible, but with different parameters, by the competent water authority and by the operator of the public wastewater facilities. For economic reasons, however, the monitoring is usually coordinated accordingly.

In China, the CJ 343-2010 discharge standard for indirectly discharged industrial wastewater has been in place for 10 years (see Table 6.4). This standard can be tightened regionally by the water authorities due to local boundary conditions. In practice, it has been found that operators often do not have sufficient information on industrial waste water and are not involved in monitoring discharges. This explains why Chinese municipal wastewater treatment plants are experiencing difficulties in meeting quality requirements.

The biggest challenge for the investigated Suzhou II wastewater treatment plant is the industrial wastewater to be treated. This is all the more true because in the future even more industrial plants will have to be connected to the treatment plant and significant production increases are expected from the plants with correspondingly higher wastewater loads. Therefore, for further operational optimisation of the treatment plant, it seems urgently necessary to establish a complete indirect discharger register in accordance with the German experience. At the moment there is only an incomplete directory. Although Huayan Water, the operator of the investigated wastewater treatment plant, is aware of the industrial enterprises connected to the treatment plant, it has no meaningful information on the fluctuations in the quality of the wastewater and the effects of in-house pretreatment.

It is strongly recommended that industrial wastewater discharges be monitored in consultation with the competent authorities. Based on German experience, the following strategy, which is useful for monitoring, should be used for this purpose:

- On-site monitoring of commercial and industrial indirect dischargers
- Monitoring of discharges at nodes in the sewer network
- Detection of illegal indirect discharges.

It is advisable to carry out online monitoring of the pre-treatment plants, at least for industrial plants with highly contaminated wastewater. In the event of disruptions in the operation of these plants and transmission of online data, measures such as commissioning emergency basins or dosing powdered activated carbon could be initiated at an early stage at the downstream wastewater treatment plant.

The monitoring of nodes in the sewer network is used to trace sporadic pollutant discharges. The regular taking of random samples or better of permanent mixed samples at selected sewer nodes is a useful supplement to the monitoring of indirect dischargers.

Different matrices can be used to detect unauthorized indirect discharges. Firstly, the information on the various industrial effluents discharged. However, an indirect determination can also be made about the nature of the biofilm on the sewer pipe walls or the sewer sediments. However, only pollutants that accumulate in solids are recorded.

6.5.2 Optimisation of the Existing Sedimentation Stage

One approach to increase nitrogen elimination in existing wastewater treatment plants is the optimization of relevant treatment facilities. In the case of the Suzhou II wastewater treatment plant under investigation, it therefore seemed sensible to carry out investigations to improve the effect of sedimentation on denitrification in the plant. The main objective was to improve the C/N ratio for nitrogen elimination in the biological stage.

In order to identify the sedimentation behaviour of the wastewater in the influent of the primary treatment plant more precisely and to demonstrate the sedimentation performance of the primary treatment plant, sedimentation tests were carried out in the treatment plant laboratory. To carry out the sedimentation tests, 5 L of wastewater were taken from the inflow of the sedimentation stage on 4 days at intervals of 2 days. Figure 6.6 shows the setup for the laboratory tests.

After reaching the respective sedimentation time (0/15/30/60/80 min) a sample was taken from the respective vessel (Imhoff cone). The parameters COD (total and solved), BOD5, suspended solids, NH4-N and TN were then analysed for the five different samples.

The evaluation of the results of the sedimentation tests with regard to the suspended solids (SS) is shown in Table 6.5. It is shown that the concentration of SS decreased as expected over the 80 min period considered. The discernible fluctuations over time can be explained by the small number of measured values.

Table 6.5 shows that after 80 min of discontinuation a reduction of the suspended solids of around 26%. This does not quite correspond to the reduction for municipal wastewater, but in view of the high proportion of industrial wastewater in wastewater treatment plant influent realistic.

A similar decrease of around 27% was observed for the COD, as shown in Table 6.6. However, the reduction of the COD did not result from the removal of the suspended COD fraction.

Table 6.6 also shows an 8% decrease in dissolved COD after a settling time of 80 min. The BOD5 has not been analysed. However, it can be assumed that a greater reduction than for the COD has been achieved. This is assumed to be 33% for the carbon–nitrogen ratio.

No major effect on nitrogen values is to be expected from sedimentation. To show this also the laboratory tests performed. Table 6.7 shows the values for the total nitrogen is listed. This results in a decrease after 80 min settling time concentration of around 16%.

For denitrification, the carbon/nitrogen ratio (C/N) in the biological treatment stage is of decisive importance. According to Tables 6.6 and 6.7, the inflow to primary treatment had an average COD/TN ratio of 11.45. Since it is not the COD but the readily biochemically available carbon that is relevant for the assessment of this ratio, a BOD5/TN ratio of 6,7 can be determined in the sedimentation inflow, based on the mean values for BOD5 and TN according to Figs. 6.4 and 6.5. At reduction rates of

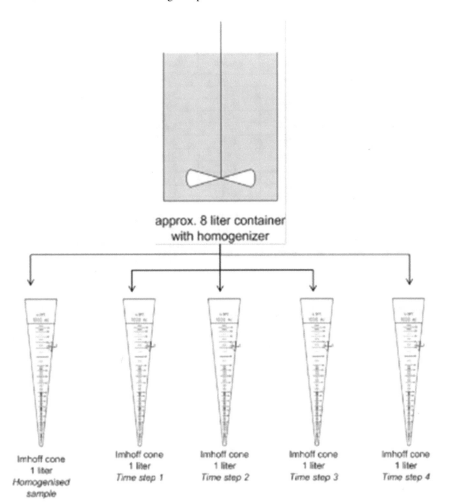

Fig. 6.6 Setup for laboratory investigations to optimize the sedimentation stage of investigated Suzhou II wastewater treatment plant

Table 6.5 Concentration of suspended solids at sedimentation investigations in the laboratory of the Suzhou II wastewater treatment plant

	settling time [min]	Minimum	Median	Maximum	85%-Quantile	Average
	0	105,0	128,5	185,0	179,0	139,8
	5	98,0	112,0	126,0	121,8	112,0
	10	85,0	103,0	121,0	115,6	103,0
SS	15	77,0	115,5	163,0	147,3	120,5
	25	75,0	88,5	102,0	98,0	88,5
	30	70,0	99,0	161,0	136,3	108,5
	60	80,0	123,5	163,0	137,8	120,7
	80	80,0	100,0	132,0	117,6	103,0

Table 6.6 COD values for sedimentation tests in the laboratory of the Suzhou II wastewater treatment plant

	settling time [min]	Minimum	Median	Maximum	85%-Quantile	Average
COD_total	0	320,0	425,0	544,0	475,0	419,3
	5	314,0	368,0	422,0	405,8	368,0
	10	304,0	355,0	406,0	390,7	355,0
	15	298,0	368,5	415,0	401,5	360,3
	25	296,0	332,5	369,0	358,1	332,5
	30	290,0	345,5	365,0	358,3	333,7
	60	292,0	308,0	348,0	337,2	314,0
	80	283,0	301,0	344,0	325,6	307,3
COD_solved	0	124,0	165,0	195,0	183,8	164,7
	5	158,0	165,5	173,0	170,8	165,5
	10	156,0	161,5	167,0	165,4	161,5
	15	121,0	155,0	181,0	169,0	155,0
	25	154,0	154,0	181,0	169,0	154,7
	30	118,0	154,5	183,0	165,0	153,0
	60	137,0	155,0	181,0	171,1	157,0
	80	132,0	149,5	177,0	168,0	152,0

Table 6.7 Total nitrogen values in sedimentation tests in the laboratory of the Suzhou II wastewater treatment plant

	settling time [min]	Minimum	Median	Maximum	85%-Quantile	Average
TN	0	32,0	36,8	40,6	38,9	36,6
	5	34,6	36,9	39,2	38,5	36,9
	10	34,1	36,4	38,6	37,9	36,4
	15	31,2	34,6	38,7	37,8	35,1
	25	33,8	35,6	37,4	36,9	35,6
	30	29,7	33,9	36,0	36,0	33,8
	60	28,8	32,5	35,8	34,8	32,4
	80	26,9	29,7	36,1	33,5	30,6

33 and 16% by sedimentation, respectively, this ratio would decrease to a 5.3 (Table 6.8 (Lee 2018)).

The above investigations of the sedimentation behaviour of the wastewater of the investigated wastewater treatment plant raise the question whether sedimentation before the biological stage should be avoided in the future. The existing sedimentation plant does have a number of advantages, such as the increased sludge age in the biological treatment stage and the use of the sludge produced in the primary sedimentation stage for energy generation in the existing digestion plant. However, the conditions for extensive denitrification deteriorate due to the reduced carbon/nitrogen ratio. The laboratory tests carried out have confirmed this.

There are several ways to reduce the negative effect of an existing sedimentation stage on denitrification. This can be done by reducing the settling time by taking a part of the tank volume out of operation. Another possibility is a partial bypass of the sedimentation, so that wastewater not settled with a partial flow is fed to the biological

Table 6.8 C/N ratio during sedimentation tests in the laboratory of the Suzhou II wastewater treatment plant

	sett-ling time [min]	Average
C/N	0	11,5
	5	10,0
	10	9,8
	15	10,3
	25	9,3
	30	9,9
	60	9,7
	80	10,0

treatment stage. In view of the very extensive nitrogen elimination intended for the Suzhou II treatment plant, it does not appear advisable in any case to increase the sedimentation effect by increasing the volume or installing lamella separators.

6.5.3 Optimisation of Nitrogen Elimination

The significantly increased requirements for nitrogen elimination in wastewater treatment at Lake Tai means for the Suzhou II wastewater treatment plant opposite previous services a more extensive denitrification. This chapter will therefore on investigations carried out for various process engineering measures to improve the denitrification performance of the treatment plant.

6.5.3.1 Optimizing the Dosage of External Carbon

The heterotrophic microorganisms responsible for denitrification obtain the carbon and their energy from external organic sources. In contrast to the autotrophic nitrificants, they are not able to synthesize their energy from oxidation reactions themselves (Görner 1999). An optimal carbon/nitrogen ratio is between 12:1 and 15:1, but most of the municipal wastewater treatment plants in China have much lower ratios. These are usually between 6:1 and 8:1, which indicates an excess of nitrogen and limits denitrification. Very low total nitrogen contents cannot be achieved with this method. However, such conditions can be improved by adding an external carbon source.

The external carbon sources that can be used can be divided into conventional external, unconventional external and internal sources. In Germany, materials that can be assigned to these three groups have proven their worth in wastewater practice.

Conventional carbon sources include substances such as methanol, ethanol, sugar beet syrup, acetic acid and technical salts. These offer different conditions. Thus, acetic acid is suitable for a rapidly effective support of denitrification, as the microorganisms require a very short adaptation time to the substrate. Methanol, on the other hand, is only converted by certain microorganisms, so that its effectiveness is lower.

In addition to conventional carbon sources, other carbon sources are increasingly gaining in popularity. These offer the advantage that they do not have to be produced separately, but are produced as waste or by-products in various industrial production processes and are usually much cheaper than conventional substrates. Examples of alternative external carbon sources include glycol-containing wastewater from chemical production, high-carbon wastewater from the beverage industry, breweries, the baking industry and ice cream production, and whey powder and crude glycerine from biodiesel production (Amed et al. 2000).

In addition to external carbon sources, there is also the possibility of using carbon-containing internal material flows in wastewater treatment plants to improve the denitrification capacity. For example, a targeted use of sewage sludge or sludge water is suitable for this purpose. A further possibility for activating carbon potential to improve denitrification is the comminution (disintegration) of the sewage sludge by mechanical or thermal means. An additional advantage of sewage sludge disintegration is the minimization of the sludge mass to be disposed of. A screenings washing also leads to an internal carbon source to be used and also reduces the disposal costs for the screenings.

Table 6.9 lists typical carbon sources for improving denitrification with their origin, advantages and disadvantages.

In the wastewater treatment at the Suzhou II wastewater treatment plant neither external carbon sources nor specific measures to use internal carbon potentials have been applied so far, since the requirements for nitrogen discharge values could be met even without such measures. Due to the need for more extensive nitrogen elimination in the future, a series of laboratory tests were carried out in the wastewater treatment plant laboratory in spring 2017 to determine the effect of a dosage of carbon on nitrogen elimination in the biological stage. Acetic acid and sodium acetate were added under anoxic conditions with different stirring times. Sodium actate is considered a preferred agent in China for improving the carbon/nitrogen ratio. Despite fluctuating results, the investigations can be described as successful.

However, for economic reasons and from the point of view of sustainability, in the long term the large-scale use of a carbon source derived from industrial residues should be aimed at. It may also be possible to use substances from industrial plants in the catchment area of the wastewater treatment plant. Once an indirect discharger register is available, the corresponding materials would have to be identified and examined in the laboratory for their suitability and effect before being used on an industrial scale.

Table 6.9 Advantages and disadvantages of different external carbon sources to improve the denitrification of wastewater

External carbon sources		
	Advantages	**Disadvantages**
Methanol	• frequently used • low cost • low sludge accumulation	• adaptation to biocenosis required • potential toxic effect
Ethanol	no adaptation to biocoenosis required	• higher sludge accumulation than with methanol • potential toxic effect • higher costs than methanol
Acetate	• no adaptation to biocoenosis required • low sludge accumulation	• high substrate demand • higher costs than methanol and ethanol
Industrial residues		
Industrial wastewater z. B. brewery wastewater	• almost free • desirable recycling product	• not constant condition • quality control required • laboratory tests required to determine the amount to be added
Internal carbon sources		
	Advantages	**Disadvantages**
acidified sewage sludge, sludge water	• low cost • no entry of foreign substances	• possible additional N-entry • not constant condition

6.5.3.2 Use of a Downstream Anoxic Filter

A possible technical solution for further nitrogen removal is the use of denitrification after the biological treatment stage. For this purpose an anoxic filter system is the best choice. At the Suzhou II wastewater treatment plant, a pilot plant designed and built in Germany was used for investigations. Figure 6.7 shows the test room with the pilot plant. The test facilities were housed in the container used to transport the plant from Germany to China.

Figure 6.8 shows a schematic diagram of the test facility with all its individual parts. The filter material used was an expanded clay material with a grain size of 4 to 8 mm and a quantity of 90 L, which has proven itself in Germany for filtration. The carbon dosage was carried out using sodium acetate solutions (CH3COONa) prepared in the laboratory. To assess the elimination performance of the pilot plant at

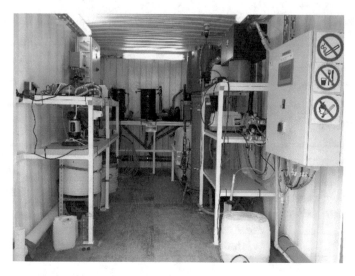

Fig. 6.7 Test room with the anoxic pilot filter system for the downstream denitrification at the Suzhou II wastewater treatment plant

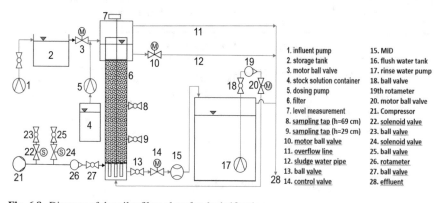

1. influent pump	15. MID
2. storage tank	16. flush water tank
3. motor ball valve	17. rinse water pump
4. stock solution container	18. ball valve
5. dosing pump	19th rotameter
6. filter	20. motor ball valve
7. level measurement	21. Compressor
8. sampling tap (h=69 cm)	22. solenoid valve
9. sampling tap (h=29 cm)	23. ball valve
10. motor ball valve	24. solenoid valve
11. overflow line	25. ball valve
12. sludge water pipe	26. rotameter
13. ball valve	27. ball valve
14. control valve	28. effluent

Fig. 6.8 Diagram of the pilot filter plant for denitrification

higher nutrient loads, an increase of nitrate in the filter influent in the form of sodium nitrate (NaNO3) was carried out in individual test sections.

The pilot tests carried out concerned the following parameter ranges:

– Nitrate addition with NaNO3 (max. 30 mg/l NO3-N),
– C:N ratio by carbon addition with CH3COONa (max. COD:TN = 5:1),
– Flow rates between 30 and 110 [l/h]

With a multiparameter probe the parameters pH, redox potential, oxygen content, conductivity, dissolved solids and temperature could be measured. In the la-bor of the wastewater treatment plant the parameters COD, TN, nitrate, nitrite, NH4-N and

TP were analysed. The BOD5 was analysed provisionally. The analytical methods were carried out according to the Chinese standards.

Table 6.10 gives an overview of the technical data of the pilot filter plant and the feed data.

The COD and nitrogen values recorded during the investigation period in the effluent of the final clarification and in the filter inlet are shown in Fig. 6.9.

During the test period, there were significant fluctuations in the characteristics of the pilot plant influent, as shown in Fig. 6.9. This was particularly true in May 2018, when a very pronounced disturbance of the operation of the treatment plant apparently occurred. The effluent values of the secondary clarification increased to a multiple of the previous values both for the COD and for the total nitrogen. It is remarkable that in spring 2018 the major part of the oxidised nitrogen in the effluent of the biological clarification stage was present as nitrite nitrogen and during the

Table 6.10 Technical data for the operation of the pilot filter plant for denitrification

Feed flow	30	60	90	110	[l/h]
Flow time	122,1	61,0	40,7	33,3	[min]
Filter velocity	1,12	2,25	3,37	4,12	[m/h] [m³/(m²·h)]
Filter material	Liaperl G 4-8 mm				[-]
Filter bed height	1,25				[m]
Filter surface	50,42				[m²]
Effective reactor volume	0,056				[m³]
Filter bed volume	0,09				[m³]
Filter material bulk density	717				[kg/m³]
Surface of the filter material	900				[m²/m³]

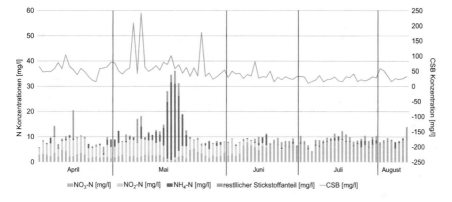

Fig. 6.9 COD and nitrogen values in the influent of the anoxic pilot filter in the test period

Table 6.11 Average influent and effluent concentrations and elimination efficiency of the pilot filter for different parameters during a test series without addition of external carbon

Parameter	n=18	COD	TN	NO₃-N	NO₂-N	NH₄-N	TP
Influent concentr. (mg/L)	Ø	26,1	8,52	6,61	0,11	0,37	0,67
	s	9,5	1,56	1,19	0,04	0,28	0,19
Effluent concentr. (mg/L)	Ø	19,67	7,3	5,99	0,2	0,6	0,33
	s	7,78	1,41	1,19	0,09	0,32	0,05
Elimination rate (%)	Ø	21,26	13,86	9,44	-1,17	-140,79	47,69
	s	24,53	9,15	8,17	1,06	209,62	15,78
COD$_{elim.}$/TN$_{elim.}$				Ø = 5,04			

disturbance of the wastewater treatment plant the nitrification in the plant came to a complete standstill at times.

As a basis for investigations with the addition of external carbon, a test series without carbon addition was run. This investigation was carried out at midsummer temperatures. The COD/TN ratio in the effluent of the secondary treatment was on average 3.0. The mean concentrations and standard deviations in the inflow and outflow as well as the elimination rates for the main parameters are summarised in Table 6.11.

The results according to Table 6.11 were obtained at very moderate inflow values, so that under such conditions the current requirements for sewage plant effluents in the Tai Lake area could be met. However, this does not apply to the very low total nitrogen value of 6 mg/L which is aimed for in the future.

The pilot filter tests with the addition of sodium acetate as an external carbon source were carried out in April 2018 at low temperatures and in July/August 2018 at summer temperatures. Table 6.12 lists the conditions for these test phases.

Table 6.12 Process conditions for the pilot filter trials with addition of sodium acetate

Parameters (n = 27)		Min	Max	Ø	s
pH value	[-]	6,35	6,81	6,53	0,16
Redox potential	[mV]	42,17	228,97	130,76	53,27
Oxygen content	[ppm]	0,08	1,1	0,47	0,39
Conductivity	[µS/cm]	1.062	1.491	1.292,1	137,79
Water temperature	[°C]	19,84	29,53	26,45	3,88
Air temperature	[°C]	8	37	27	6,31
COD/TN ratio	[-]	3,1	13,1	7,34	2,53
COD filter bed load	[kg/(m³·d)]	0,38	2,23	1,11	0,49
TN filter bed load	[kg/(m³·d)]	0,06	0,36	0,18	0,09
NO3-N filter bed load	[kg/(m³·d)]	0,02	0,25	0,11	0,07

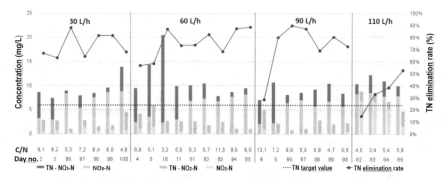

Fig. 6.10 Nitrogen values in the influent and effluent of the pilot filter and C/N values and TN elimination rate for test period with addition of sodium acetate

In April 2018, high nitrite contents of the total nitrogen were present in the effluent from the secondary clarification, while the higher temperatures in summer 2018 led to improved nitrate formation. Figure 6.10 summarizes the results of the test series with carbon addition. The blue bars represent the influent values and the beige bars the effluent values.

Figure 6.10 also shows the different feed rates of the filter system. Even with very high total nitrogen values and a feed rate of up to 90 L/h, the target effluent value of 6 mg/L could be maintained. This was no longer possible consistently at the feed rate of 110 L/h. However, such a high charging rate allows the effluent value of 10 mg/L, which will apply to the Tai Lake area from 2021, to be maintained.

In the investigations on downstream denitrification, compliance with the required or desired COD effluent values was of particular interest. Figure 6.11 shows the COD influent and effluent values according to the nitrogen values shown in Fig. 6.10. The beige bars represent the influent values and the grey bars the effluent values.

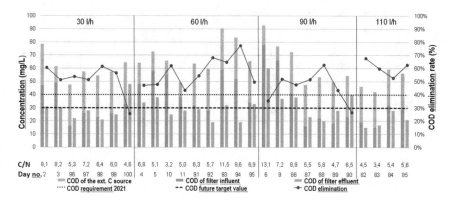

Fig. 6.11 COD values in the influent and effluent of the pilot filter, C/N values and COD elimination rate for test period with addition of sodium acetate

Only two of the COD effluent concentrations recorded were above the limit of 40 mg/L applicable to the Tai lake area from 2021, but 40% of the recorded one above the long-term target of 30 mg/L. The main reason for these exceedances is the high non-degradable or poorly degradable COD content in the secondary treatment effluent, which is not uncommon for industrial wastewater. A future permanent compliance with COD discharge values not exceeding 30 mg/L will require improved monitoring of industrial discharges with high shear-degradable COD concentrations. Indeed, the results shown in Fig. 6.11 make it clear that the high COD effluent values of the secondary treatment were responsible for the insufficient COD effluent values of the filter and not the carbon added externally before the filtration stage.

Because the elimination rates for COD and total nitrogen were relatively low in the first week of the experiments with carbon addition with increased feeding of the pilot filter, it was decided to carry out a series of experiments with additional nitrate addition. The discharge and inflow values as well as the elimination rates for nitrogen are shown in Fig. 6.12. The blue bars represent the influent values and the beige bars the effluent values.

The outside temperatures during this series of tests averaged 24 °C. Even with the addition of sodium nitrate, relatively low influent concentrations for the total nitrogen were obtained. According to Fig. 6.12, the effluent values were clearly below the future target total nitrogen value of 6 mg/L even at a feed rate of 90 L/h.

The influent and effluent values for COD and COD elimination are shown in Fig. 6.13. The beige bars represent the influent values and the grey bars the effluent values.

On all days of the study, the COD effluent value of 40 mg/L, which will apply from 2021 for wastewater treatment plants in the Tai Lake area, was maintained regardless of the filter feed and despite the carbon dosage. With the exception of one day, this also applied to the future target value of 30 mg/L.

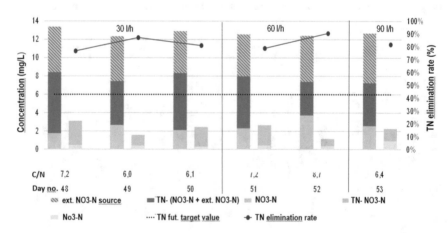

Fig. 6.12 Influent and effluent concentrations for TN and NO3-N and TN elimination rate for test period with addition of external carbon and nitrate

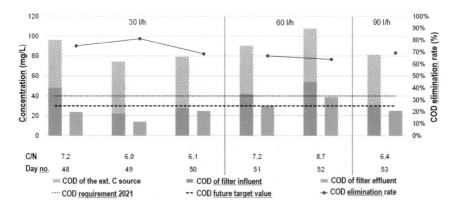

Fig. 6.13 COD values in the influent and effluent of the pilot filter, C/N values and COD elimination rate for test period with addition of sodium acetate and nitrate

The results of all investigations with the anoxic pilot filter system can be summarized as The results of all investigations with the anoxic pilot filter system can be summarised as follows:

The purpose of the pilot tests was to improve the nitrogen elimination taking place in the biological treatment stage down to very low total nitrogen concentrations. As the COD in the effluent of the final clarification was, as expected, only available to a small extent as a substrate for denitrification, the addition of an external carbon substrate proved to be very effective. Very low total nitrogen values could be achieved, so that the future target value of 6 mg/L at most would have to be met. However, the prerequisite for such performance of the downstream denitrification stage is a functioning nitrification in the upstream biological treatment stage. The massive disturbance of the wastewater treatment plant during the test period, presumably caused by toxic substances in the industrial wastewater, led to an extensive prevention of nitrification. Reliable online monitoring and technical measures in the indirect discharger area or at the Suzhou II treatment plant must prevent such disturbances if permanent and extensive nitrogen elimination is to be guaranteed in future.

Compliance with future COD requirements in the effluent of the treatment plant will not depend on the addition of an external carbon source. Given the existing process engineering situation of the wastewater treatment plant, it is necessary that the inert COD content in the effluent of the secondary clarification will in future be below the required COD effluent values of 40 or 30 mg/L. If this cannot be achieved due to the industrial wastewater treated in the treatment plant, a specific elimination stage for the hardly degradable organic wastewater constituents would have to be provided.

6.5.3.3   Sludge Water Treatment

As mentioned in Sect. 6.4, investigations were also carried out at the Suzhou WWTP using pilot plants for nitrogen elimination from the sludge water substream. The aim is not to improve the elimination performance by treatment in the main wastewater treatment stream, but in the sludge treatment substream. In Germany, corresponding process engineering solutions have already proven themselves on a large scale.

Operation of an SBR-MBBR Experimental Plant
The investigations were carried out in the summer months of 2018 and were carried out by the German company EvU. The core of the plant was a Moving Bed Biofilm Reactor (MBBR), which was operated as a sequencing batch reactor. The test plant, which is located in a container at the Suzhou II wastewater treatment plant, is shown in Fig. 6.14. It consisted of a storage tank, an SBR reactor and an outlet tank for sampling as well as a switchgear for control. The reactor had a diameter of 30 cm and a height of 51 cm corresponding to a volume of 36 L. EvU-Pearl® plastic bodies with a specific surface of 700 m^2/m^3 served as carrier material. The 10.8 L of material used corresponded to a filling ratio of 30%.

The MBBR plant was fed with the process water of the mechanical sludge dewatering of the sewage treatment plant in order to determine its biodegradability with regard to nitrogen and COD.

The storage tank was used to store the wastewater to be treated. The wastewater was treated in batches according to a specified cycle. By switching off the aeration the cleaning process of the pre-cycle was finished and the sedimentation phase was initiated. The sedimenting carrier material EvU®-Pearl settled at the bottom of the SBR. Above the sedimentation layer a clear phase with purified waste water formed. The sedimentation phase was followed by the withdrawal phase. Excess sludge was

Fig. 6.14   MBBR pilot plant at Suzhou II wastewater treatment plant for sludge water treatment

Table 6.13 Cycle settings during the entire operating period of the MBBR pilot plant

Versuchsphasen	I	II	III	IV	V	VI	VII	VIII
Phasen/Zeit(s)	ab 04.06	ab 22.06	ab 16.07	ab 24.07	ab 30.07	ab 10.08	ab 27.08	ab 27.09
Füllen	60	60	60	60	60	60	60	60
Stillstand	100	100	100	100	100	100	100	0
Mischen	500	800	1160	520	1640	1730	1630	1515
Belüften	1200	900	540	780	1090	1200	1200	1515
Sedimentieren	3180	3180	3180	1980	4200	3000	3600	3000
Abzug	60	60	60	60	60	60	60	60
Gesamtzyklus	14400	14400	14400	10800	21600	21600	21600	21600

removed at specified intervals and fed into a sludge storage tank. Four to eight cycles were run daily.

After a two-week running-in phase, eight test phases were carried out in the investigation period from June to September 2018. Table 6.13 shows the test periods and the SBR cycles for these test phases.

Due to low denitrification performance in the initial test phases, the dosing of an external one was later carried out according to the following schedule.

CSB-Zugabe (mg/L):	ab 09.07.	ab 06.08.	ab 20.08.	ab 27.08.	ab 03.09.	ab 12.09.
	100	150	160	170	180	190

The investigations concentrated on nitrification and especially on nitrogen elimination. Of course, the influence of the feeding cycles was also of importance. Figure 6.15 shows the inlet and outlet concentrations of ammonium nitrogen and the corresponding elimination rates during the eight investigation phases applied. It should be noted that the concentrations of ammonium nitrogen in the sludge water in Germany are several times higher than the concentrations shown in Fig. 6.15 and range between 30 and 50 mg/L.

In the middle of phase I and VII, good NH4-N oxidation was temporarily achieved at an aeration ratio of 67%/33%. As expected, the elimination performance decreased with reduction of the aeration time. In total, more than 50% of the ammonium could be oxidized continuously.

Nitrogen elimination was very low in the first four test phases, as can be seen from Fig. 6.16. The reason for this was the relatively low COD with values between 50 and 200 mg/L, most of which is not available for the necessary denitrification. This changed with increased addition of external carbon.

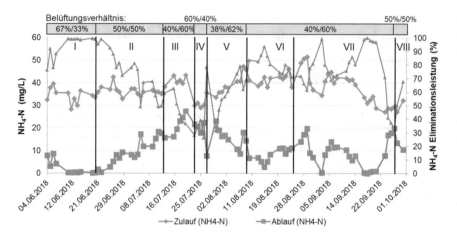

Fig. 6.15 Inflow and effluent concentrations of NH4-N and elimination performance as a function of cycle setting and aeration ratio during the investigations with the MBBR pilot plant

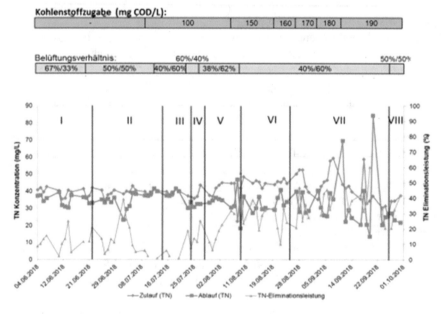

Fig. 6.16 Inflow and effluent concentrations of total nitrogen TN and elimination rates during the investigations with the MBBR pilot plant

Figure 6.16 shows the inflow and outflow values for total nitrogen and the corresponding elimination performance. The respective aeration ratio and the addition of carbon were also indicated. The moderate elimination rates for the nitrogen are also clear from this picture.

Investigations on the correlation between the ammonia nitrogen surface load of the carrier bodies and the surface oxidation performance showed that surface loads between 4.5 and 6 g/m^2 * d led to good performances. These values could be used as a basis for plant dimensioning in the event of a possible large-scale technical application.

In summary, the tests carried out confirmed that the nitrogen loads present in the process water of mechanical sludge dewatering can be reduced by means of the MBBR process. Thus the main biological load can be reduced accordingly. However, it appears questionable whether such a bypass treatment stage is economically viable at the low ammonia nitrogen concentrations and COD values, since an acceptable nitrogen elimination performance can only be achieved with the aid of an external carbon source.

Due to the low ammonium concentrations of the sludge water, the investigations also did not include the installation of a magnesium-ammonium-phosphate separation (MAP stage). Such a process engineering solution could otherwise have been installed in the feed tank of the test plant and would have ensured a desirable removal of ammonium from the sludge water.

Operation of a Pilot Plant for Deammonification
The sludge water resulting from the dewatering of digested sludge is highly nitrogenous and is usually returned to the biological wastewater treatment plant. This means an additional load of about 10 to 15% of the nitrogen input load and makes a separate treatment of the sludge water appear sensible because the process stability of nitrogen elimination can be significantly improved. In recent years, so-called deammonification has proved to be an effective method of such a bypass treatment. Deammonification in particular, with its high energy and cost saving potential, has currently led to an increased number of partial flow plants being built in practice. At present, approx. 70 plants for the treatment of sludge water in partial flow are operated in Germany.

In the process of deammonification, about half of the ammonium contained in the sludge water is first converted to nitrite under aerobic conditions, which is then converted to gaseous nitrogen and about 10% nitrate under anaerobic conditions with the help of specialized bacteria (planktomycetes) together with the other half of the ammonium. Figure 6.17 shows the basic principles of conventional nitrification/denitrification and deammonification (Lackner et al. 2013).

Compared to the classic nitrification/denitrification, the ammonia ionization process uses 60% less oxygen and 100% less carbon. Furthermore, due to the slow growth rates of the microorganisms involved, considerably less excess sludge is produced.

To demonstrate deammonification in China, a laboratory test plant was operated at the Suzhou II wastewater treatment plant in 2018. The German company ATEMIS, which has extensive practical experience in deammonification, was responsible for this. The sewage treatment plant was suitable for the investigations because it has a sewage sludge digestion system and a similarly high amonium nitrogen load in the sludge water as in Germany was expected. It turned out that for the investigations

Nitrification/Denitrification   Deammonification

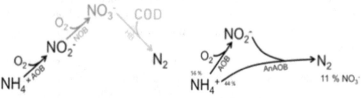

Fig. 6.17 Comparison of conventional nitrogen elimination and deammonification of wastewater

an increase by external ammonium nitrogen was necessary due to the very low contamination of the sludge water on site.

The functionality of the experimental plant could not be fully exploited because the very specific sludge required for nitritation was not available in sufficient quantity. Since such a sludge did not exist in China up to now, a transport of corresponding quantities of inoculated sludge from Germany was planned for the start-up of the pilot plant. In order to avoid difficulties during a long lasting transport in shipping containers or air freight, a freeze drying of suitable inoculation sludge was carried out in Germany and the sludge was taken to China as hand luggage. This measure was basically successful, as the dried inoculation sludge could be revived in the test plant. However, due to the limited drying capacity, the amount of sludge and the bacterial mass was too small for full-load operation of the test plant.

For economic reasons, the deammonification of nitrogen-contaminated sludge water will only be feasible in China if the nitrogen potential to be oxidized is at least 100 mg NH4-N. In any case, preliminary practical tests should be carried out.

6.6 Recommendations for Future Wastewater Treatment in the Investigated Wastewater Treatment Plant

The background to the investigations carried out at the Suzhou II wastewater treatment plant in 2017 and 2018 was the finding that improved treatment performance of the plant appears necessary to meet future requirements. The following recommendations for the future design and operation of the plant can be derived from the investigations.

The operation of wastewater treatment plants with a high proportion of industrial wastewater requires precise knowledge of the industrial wastewater to be treated. This also includes immediate information in case of operational disturbances or changes in production in the industrial wastewater producing companies, if these have an impact on the wastewater quality. The operator of the wastewater treatment plant under investigation did not have any complete information on this subject until now. It is recommended to establish a regularly updated cadastre for the industrial companies connected to the treatment plant, including all relevant wastewater data including the wastewater pretreatment plants. With regard to operational pre-treatment measures, the competent supervisory authority should clarify whether a targeted partial flow treatment of particularly polluted wastewater should be carried out instead of pre-treating the entire operational wastewater.

In future, compliance with very low nitrogen levels will mean targeted measures for the Suzhou II treatment plant. One possibility is the dosing of external carbon for denitrification in the biological treatment stage. For economic reasons and sustainability, it is recommended that suitable organic residues from operations in the Suzhou Industrial Park be used. However, their suitability and effect would have to be investigated in the laboratory before their large-scale use.

In view of the necessity to maintain very low effluent values for nitrogen in the future, the use of a denitrification in an anoxic filter plant downstream of the biological treatment stage seems very sensible. Due to the low availability of COD in the effluent of the final clarification to ensure nitrate respiration in the filter plant, the addition of an external carbon substrate will become necessary. To adapt to the purification requirements for compliance with the requirements, it may make sense for economic reasons to lead only a partial flow of the secondary clarification effluent through the filter plant, at least temporarily.

In order to meet the requirements of the COD effluent values of the treatment plant, it will be necessary in the future that the inert COD content in the final sedimentation effluent is below the required COD effluent values of 40 mg/L or 30 mg/L. If this cannot be achieved due to the industrial waste water treated in the treatment plant, a specific elimination stage must be provided for the hardly degradable organic waste water constituents. For this purpose, either adsorption with activated carbon or chemical oxidation is suitable.

6.7   Summary

The special importance of Lake Tai for the water supply in the Shanghai area has led to very demanding new requirements for wastewater treatment in the catchment area of the lake, due to the high levels of pollution, especially of plant nutrients, that have been known for years. Within the scope of German research activities belonging to the Chinese Mayor Water Programme, a contact was established with the Chinese company Huayan Water Affairs Co., Ltd., which is responsible for the water supply and wastewater disposal in the Suzhou Industrial Park on Lake Tai. This resulted in a research project financed by the German Ministry of Education and Research (BMBF) and the company Huayan Water. Using the example of the Suzhou Sewage Treatment Plant II in the Industrial Park Suzhou, which has to treat a considerable proportion of industrial wastewater, approaches for the future improvement of wastewater treatment were to be investigated on the basis of German experience.

The focus was on extreme demands on the wastewater treatment plant effluent with 6 mg/L total nitrogen and 30 mg/L COD.

The investigations carried out covered both technological and management aspects. For the future successful function of the investigated wastewater treatment plant it seems to be important that all basic and current data and information about the different wastewaters of the connected industrial plants are available during the treatment operation.

Investigations in the laboratory of the sewage treatment plant and with the help of three pilot plants in operation resulted in various findings. This has resulted in the following suggestions and notes, among others.

- It seems worth considering whether in future the sedimentation plant upstream of the biological treatment stage should continue to be used as such or whether it should be used for other purposes.
- For the dosing of an external carbon source that will be necessary in the future to improve nitrogen elimination, suitable organic substrates from the Suzhou Industrial Park should be used as industrial residues instead of conventional products.
- For the stability of a future very extensive denitrification the arrangement of an anoxic filtration plant downstream of the biological clarification stage with dosing of external carbon seems to be reasonable.
- If, in future, it will not be possible to permanently reduce the COD that is difficult to degrade during the final clarification process, an additional procedural measure will be necessary. Either activated carbon adsorption or chemical oxidation can be considered for this purpose.

In conclusion, it should be noted that the research project carried out resulted in a very trustful and harmonious cooperation between all Chinese and German partners involved.

References

Amed J, Bumiller W, Kusche I, Donnert D (2000) Effiziente Abwasserreinigung durch einfache Prozessmodifikation und Nutzung von Küchenabfällen, Forschungszentrum Karlsruhe Technik und Umwelt, Karlsruhe

ABWV (2004) Verordnung über Anforderungen an das Einleiten von Abwasser in Gewässer - Abwasserverordnung – AbwV, 17.06.2004 (BGBl. I S. 1109, S. 2625)

Dohmann M (2014) Combined or separate treatment of municipal and industrial wastewater, Presentation on 11 October 2012 at the Chin Research Acad of Envir Sciences in Beijing

DWA (2013) DWA Merkblatt M 115-1 und -2 Indirekteinleitung nicht häuslichen Abwassers, Hennef

Görner K, Hübner K (1999) Umweltschutztechnik, Springer Verlag, Berlin

Lackner S, Horn H, Schreff D (2013) Pilotvorhaben Deammonifikation, DVGW, Karlsruhe

Lee V (2018) Advanced post-treatment nitrogen elimination using an anoxic biofilter on the example of a Chinese industrial wastewater treatment plant, Master thesis at the RWTH Aachen University, Aachen, Sept. 2018

Xiang L (2017) Treatment technology of pharmaceutical wastewater, SINOWATER Workshop, September 8, 2017, Bejing
Xu Y (2016) 8 Things about recycling water, China Water Risk, October 19, 2016

Chapter 7
Adaption and Introduction of German Sewer Inspection Technology in China

Marc Jansen, Jan Echterhoff, Tobias Jöckel, and Sven Sturhann

7.1 Introduction

In China in particular, cities represent the spatial concentration points of society and are always a mirror image of the macro-social development trend. Often, it is at the urban level that global megatrends firstly reveal their concrete impact on people. Whether it be rapidly increasing population growth, processes of climatic change or enormous economic growth, the consequences are pervasive and always have a profound impact on urban society and, above all, on urban systems. This is clearly evident when looking at the current drainage situation in Chinese cities. While the above-ground infrastructures have been continuously adapted to the ever-increasing economic growth in the People's Republic, the expansion and maintenance of the under-ground sewage system has been largely neglected. The consequences are devastating and the pressure to act is indispensable.

The objective of SINO-INSPECTION is to provide an important contribution to the maintenance and improvement of the drainage situation in Chinese cities. This project is realized by developing and demonstrating German sewer inspection technologies, adapting these technologies to Chinese conditions and providing scientific proof of their functional efficiency. The joint project SINO-INSPECTION focuses on the Chinese megalopolis of Jiaxing in the Zhejiang province. The practical activities and scientific verification, as well as the derivation of criteria for sewer condition

M. Jansen (✉) · J. Echterhoff
Research Institute for Water and Waste Management at the RWTH Aachen University (FiW) e. V.,
Aachen, Germany
e-mail: jansen@fiw.rwth-aachen.de

T. Jöckel
JT Elektronik GmbH in Lindau, Lindau, Germany

S. Sturhann
Bluemetric Software GmbH in Griesheim, Griesheim, Germany

© The Author(s) 2022
M. Dohmann et al. (eds.), *Chinese Water Systems*, Terrestrial Environmental Sciences,
https://doi.org/10.1007/978-3-030-80234-9_7

assessment as part of SINO-INSPECTION, were supported by the urban drainage companies in Jiaxing and the Zhejiang University in Hangzhou.

The German project consortium consists of employees of the Research Institute for Water and Waste Management at RWTH Aachen University (FiW), employees of the company JT-elektronik GmbH and employees of the company bluemetric software GmbH.

7.2 Stage of Urban Drainage Systems in China

7.2.1 Definition of Task

Floods and odour nuisance due to massive sewer deposits as well as increased water-polluting overflows in mixed sewer systems in inner-city areas have meanwhile led to increased measures for the operation of sewer networks. This includes an increased systematic cleaning of the sewers. A continuing problem of many Chinese sewer systems and sewage treatment plants is the inflow of extraneous water. The main reason for this is the infiltration of groundwater into the sewers caused by leaking sewers, which leads to a significant dilution of the untreated wastewater. As a consequence of this extraneous water, hydraulically overloaded sewers and wastewater treatment plants are the result and, thus, higher substance pollution of surface waters is caused. Despite the above described obvious operating problems, the conditions of the existing sewers are hardly recorded, so that no targeted countermeasures can be taken. Therefore, an efficient assessment of the condition of a sewer system by means of a sewer inspection must be an important and integral part of sustainable wastewater disposal. In addition to maintaining verification of the sewer function, the inspection and the sewer maintenance based on such an inspection play a further economic role, for example as protection against flooding and consecutive damages. The structural components of the sewer network are municipal capital goods with useful lives of up to 100 years. It is thus important to preserve the substance value of the sewer system for a long time. Accordingly, a long preservation of the substance value of the sewerage system is relevant. The methods of optical inspection by means of TV cameras, which have proven their worth in Germany especially for sewers that are not accessible to the public, cannot simply be used in China. A particular difficulty in the condition assessment of many Chinese sewers is caused by the lack of possibilities to shut off the sewage flow during a TV camera inspection. Under the described boundary conditions in China and the desire there for the simplest possible inspection procedures, innovative inspection systems adapted to the conditions in China are required. The aim of the project described below is to make an important contribution to the maintenance and improvement of the drainage situation in China through the development and demonstration of German sewer inspection technologies, the adaption of the technologies to Chinese conditions and the scientific verification of their functional efficiency. Already existing contacts with the

Chinese city of Jiaxing, which has as one of the first 16 Chinese sponge cities to have a pilot character, revealed various challenges for the operation and maintenance of the sewage systems.

7.2.2   Conditions Under Which the Project Was Carried Out

The city Jiaxing is located in the north of the Chinese province Zhejiang. Jiaxing has about 3.53 million inhabitants (as of 2011) and stretches over a total area of 3,915 km².

Jiaxing is surrounded by the cities of Suzhou, Shanghai and Hangzhou, each about 100 km to the north, north-east and south-west, and is thus located in one of the more densely populated regions of the world, the Yangtze Delta region. Jiaxing itself consists of two districts, Nanhu and Xiuzhou, two counties, Jiashan and Haiyan, and three independent cities, Pinghu, Haining and Tongxiang. The morphology of the city can be considered flat, with an average altitude of 3.7 m above sea level and an average gradient of less than 1%. A large number of standing waters and rivers as well as canals are part of the city of Jiaxing. In China's eastern coastal regions, the groundwater level is on average 1–2 m below ground level and thus usually at the level of the sewer system. This leads to groundwater intrusions in large parts of the sewer system. In Guangzhou, the groundwater inflow into the sewer system is so considerable that some sewage treatment plants dilute the COD inflow to below 100 mg/l. In Zhejiang Province, 24.8% of all sewage treatment plants ave COD inflow values below 200 mg/l [11–14].

The climate is characterised by significant variations in precipitation over the course of the year. The summer months from June to August show significantly higher precipitation than the autumn/ winter months from October to December (Fig. 7.1). The annual precipitation is about 1,100 mm. Between the years 2000 and 2012, the lowest-precipitation month recorded just under 40 mm and the highest-precipitation month an average of just under 170 mm [9].

The current state of knowledge about the condition and general data of the sewage networks in China is poor and largely incomplete. Apart from overall statistics on the total sewer length in China, more comprehensive data is limited to individual investigations in only a few regions of China.[1]

A sewer system with a total length of 510,000 km exists throughout China. These consist of 210,000 km of wastewater sewers, 190,000 km of stormwater and 110,000 km of combined sewer system (status 2014, [11–14]). By way of comparison, in 2013 Germany had a total sewer system length of 575,580 km consisting of 206,234 km of wastewater sewers, 126,480 km of stormwater sewers and 242,866 km of combined sewer systems [10].

---

[1] E.g..: Huang, Dong-Bin, et al. "Confronting limitations: new solutions required for urban water management in Kunming City." Journal of Environmental Management 84.1 (2007): 49–61.

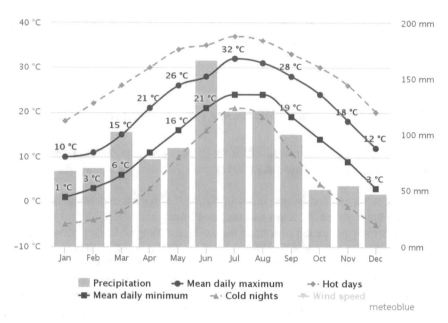

Fig. 7.1 Climate Diagram Jiaxing [9]

Within the Eco-Sponge-City demonstration area in the city of Jiaxing there are 47 km of wastewater, 140 km of stormwater and 52 km of combined sewers - this corresponds to about one tenth of the sewer length of the German city of Cologne. Almost nothing is known about the number and regularity of sewer condition. However, these can generally be classified as low. For example, in the city of Chongqing with its 10 million inhabitants, less than 5% of the entire sewer system has been inspected in the last 10 years [11–14].

The sewer damages leading to extraneous water intrusions are often caused by damage to the sewer pipe itself, leaks in the couplings or coupling offset, subsidence of individual sewer sections due to inadequate sewer bedding, or faulty connections in the separation system. These sewer damages not only lead to constructional risks within the road area and allow for contamination of the soil with wastewater, but also exert a considerable influence on the treatment performance of the connected wastewater treatment plants. The significantly increased water volume generates a hydraulic surplus resp. overloading of the wastewater treatment plants, so that the wastewater loads are no longer sufficiently reduced.

7.2.3   Present State of Science and Technology

7.2.3.1   Legal Situation

Germany

In Germany, there are a large number of different laws, regulations and standards which prescribe a certain course of action, specify a construction or operating method or serve as a guideline without direct legal binding. In the first place in the hierarchy are legal obligations as a restriction and specification for the construction and operation of drainage systems. These must be fulfilled in order to achieve legal compliance. In addition to the applicable laws, standards also have a significant influence on the construction and operation of drainage systems. Standards, according to the German Institute for Standardization (DIN), or their European (CEN) and international (ISO) pendants, do not, however, have any obligation to be fulfilled. They serve as voluntary guidelines for orientation to the current state of technology, with the exception of explicit regulations for compliance with laws or contracts (German Institute for Standardization 2017).

Legally, a drainage system is governed by the Water Resources Act (WRA) of 2009. Chapter 3, Sect. 2, § 54–§ 61 regulates the legal framework for handling wastewater.

§ 60, Sect. 7.1 states:

"Abwasseranlagen sind so zu errichten, zu betreiben und zu unterhalten, dass die Anforderungen an die Abwasserbeseitigung eingehalten werden. Im Übrigen müssen Abwasserbehandlungsanlagen [...] nach dem Stand der Technik, andere Abwasseranlagen nach den allgemein anerkannten Regeln der Technik errichtet, betrieben und unterhalten werden." (Wastewater systems must be constructed as well as operated and maintained in such a way that the requirements for wastewater disposal are met. Furthermore, wastewater treatment plants [...] must be constructed, operated and maintained in accordance with the state of the art, other wastewater plants in accordance with the generally recognised rules of technology.)

This is considered the basis for German regulations and standards applied in practice, which, as the current state of technology, are available uniformly throughout Germany as guidelines for compliance with this law. Although their application does not offer any legal guaranty of correctness of the construction and operation of drainage systems, they can help in the event of possible liability, as they make it easier to prove duly behaviour [1].

On federal level, wastewater disposal is covered by the Water Resources Act, as described above. Furthermore, § 57 states:

"Abwasser ist von den juristischen Personen des öffentlichen Rechts zu beseitigen, die nach Landesrecht hierzu verpflichtet sind (Abwasserbeseitigungspflichtige). Die Länder können bestimmen, unter welchen Voraussetzungen die Abwasserbeseitigung anderen als den in Satz 1 genannten Abwasserbeseitigungspflichtigen obliegt. Die zur Abwasserbeseitigung Verpflichteten können sich zur Erfüllung ihrer Pflichten Dritter bedienen." (Wastewater is to be disposed of by corporate bodies under public law, which are obliged to do so under federal state law (wastewater disposal obligors). The states can determine the conditions

under which wastewater disposal resides with persons other than the wastewater disposal obligors mentioned in sentence 1. Persons obliged to dispose of wastewater water may use third parties to fulfil their obligations.)

The responsibility for wastewater disposal is thus transferred to the municipalities or local authorities, with the addition that the state government also has a say. In Germany, this is achieved by means of various wastewater regulations of the state governments. It is also possible for several municipalities to transfer their duty to private service companies. These are often special-law wastewater associations that act as public-law institutions and remain as instruments of the municipalities [7].

The public sewage system, as a pure infrastructure, is thus owned by the municipalities (wastewater disposal obligors). They also take care of its structural and operational condition. Consequently, sewer construction, cleaning and monitoring as well as further steps are the responsibility of the municipalities. These appoint private, certified service providers for a sewer inspection. Larger municipalities usually have their own cleaning and inspection team.

In Europe, the DIN EN 13,508–2 for the assessment and monitoring of drainage systems exists for "the inspection and assessment of drainage systems outside of buildings part 2: Coding system for optical inspection", which is illustrated with the leaflet DWA M 149–2 of the same name by the DWA and supplements the European standard with additional specifications practiced in Germany. The two standards define a coding system for the description of observations made during an optical inspection inside sewers, manholes and inspection openings. In addition, Germany has the leaflet DWA-M 149–5 "Condition assessment and evaluation of drainage systems outside of buildings part 5: Optical inspection", which gives recommendations and assistance for the solution of technical and operational problems as well as for quality management. The aim is to ensure a qualified recording of the actual condition of the drainage systems. The process of optical inspection thus consists of the following sub-steps:

1. work preparation
2. image recording
3. image evaluation with condition description
4. documentation.

China

While a uniform Europe-wide standard exists with the realization of DIN EN 13,508–2 in Germany, in China there are various regional specifications for the assessment of the condition of drainage systems. These partly result from the fact that the official responsibility for water supply and wastewater disposal was previously not clearly regulated. This resulted in conflicts of interest regarding water use and the protection of soil and water to the disadvantage of the expansion of drainage facilities.

Since the reform from 2003 to 2007, the assessment of the condition of drainage systems has been the responsibility of the cities resp. the corresponding department for water management.

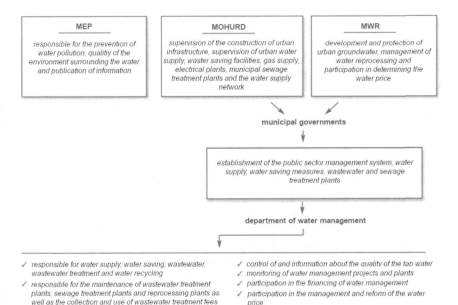

MEP	MOHURD	MWR
responsible for the prevention of water pollution, qualitiy of the environment surrounding the water and publication of information	supervision of the construction of urban infrastructure, supervision of urban water supply, waster saving facilities, gas supply, electrical plants, municipal sewage treatment plants and the water supply network	development and protection of urban groundwater, management of water reprocessing and participation in determining the water price

municipal governments

establishment of the public sector management system, water supply, water saving measures, wastewater and sewage treatment plants

department of water management

✓ responsible for water supply, water saving, wastewater, wastewater treatment and water recycling
✓ responsible for the maintenance of wastewater treatment plants, sewage treatment plants and reprocessing plants as well as the collection and use of wastewater treatment fees

✓ control of and information about the quality of the tap water
✓ monitoring of water management projects and plants
✓ participation in the financing of water management
✓ participation in the management and reform of the water price

Fig. 7.2 Management of urban water management in China according to Xinjiang conservation fund (2008)

The influences and responsibilities are shown in Fig. 7.2. This reform introduced the concept of "water management" for the first time. It combines traditional water management with water supply and wastewater disposal resp. treatment in sustainable use.

The three most important ministries responsible for monitoring the Water Management Department are the "Ministry of Housing and Urban–Rural Development" (MOHURD), the "Ministry of Environmental Protection" (MEP) and the "Ministry of Water Resources" (MWR).

As can be seen in Fig. 7.2, the drainage systems are the property of the cities, and their construction is subject to or supervised by the MOHURD. Maintenance such as sewer cleaning assessment and monitoring is the responsibility of the Department of Water Management of the city governments. The private companies are responsible for the operation of the water supply and drainage.

Due to this responsibility situation, there are currently many different sets of regulations in the various cities and company structures, which apply to the assessment of sewer condition and all further steps. In the greater Shanghai area and in the neighbouring Zhejiang province, the Chinese standard CJJ 181–2012 - "Recording and technical data of the evaluation of urban sewage systems" is common.

7.2.3.2 Recording of Sewer Condition – Hardware

The standard for the recording and assessment of the condition of drainage systems is optical inspection. This procedure generally distinguishes between direct optical inspection and indirect optical inspection.

A direct optical inspection is the inspection of manholes and accessible sewers by walking through the canal/ sewer, whereas the indirect inspection is an optical inspection by camera.

Depending on the methods used, optical inspection can essentially be used to detect and qualitatively assess clearly visible damage such as cracks, shattering, pipe breakage and collapse. Leaks without visible damage can only be detected by means of infiltrations, increased water flow at times when the volume of wastewater is normally low or when the accumulation of sediment is increased. In all other cases, leaks can only be detected by means of leak tests (cf. Annex A-2.5).

In Germany, the condition recording is preceded by a cleaning process, which has an additional positive effect. Cleaning the sewers removes blockages, which reduces sewer flooding and odour emissions (LfU 2016). Consequently, the condition recording also has a practical aspect that benefits the residents. Furthermore, there is a risk of the formation of explosive mixtures or harmful gases in the sewer system, especially in the manholes, due to putrefactive processes [2]. Regular sewer cleaning and condition assessment also limit this gas formation, as deposits and limitations of the hydraulic capacity can be eliminated at an early stage.

Innovative Procedures for Sewer Inspection and Sewer Assessment
The techniques used in Germany for optical assessment are based on a variety of camera systems. They are either carried through the sewer by wheel-based inspection crawler cameras using a flushing nozzle or they are manually pushed through the sewer. In the following two selected systems are presented, which offer the possibility to be adapted to the requirements of the Chinese sewers.

Sewer Inspection Using Electronic Sewer Mirrors (FastPicture Method)
The electronic sewer mirror (Fig. 7.3) is a camera system with integrated lighting equipment.

The example of the FastPicture system from JT-Elektronik is a Full-HD camera (resolution 1920 × 1080, zoom 360× (30× optical/12× digital)). By using highly efficient LEDs, an illumination of over 100 m is possible.

This camera system is attached to a telescopic rod, which ensures variable use with differing shaft depths. The system is installed directly at the entrance of the manhole, which means that it is no longer necessary for inspection personnel to enter the shaft.

The installation itself proves to be very simple, which results in a quick status check. The camera/ lighting unit itself has an electronic tilt mechanism, which allows the technician to manually align the camera to the optimal position for image. For the recording, the camera is zoomed into the position to be examined. The recording itself can be saved either as a photo or a video. In addition, the technician can make

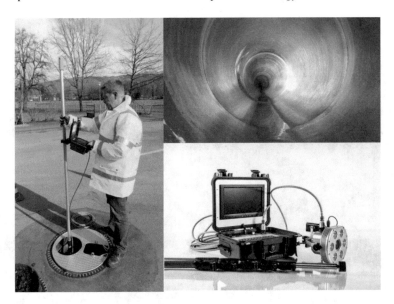

Fig. 7.3 FastPicture system by JT-elektronik GmbH (JT Elektronik)

real-time assessments via a high-resolution monitor to manually change the setting if necessary.

The use of an electronic sewer mirror does not require any prior cleaning of the area to be examined. The electronic sewer mirror can be used up to a filling level of approx. 20%.

A disadvantage of this inspection method is that damage in the sewer can only be viewed from a distance. It is not possible to make a recording for the purpose of qualitative damage assessment from a direct proximity using an electronic sewer mirror, apart from damage to the manhole itself.

Furthermore, it is possible as part of the inspection of a section of a sewer to visually inspect the condition of the shaft when lowering and raising the equipment.

In combination with inspection devices that have to be guided through the sewer, the sewer mirror enables a quick pre-assessment. Obstacles in the sewer that the inspection device cannot overcome or damage can be detected before driving through it.

Sewer Assessment Using Satellite Camera (SKI Method)

Sewer condition assessment by means of a satellite camera combines sewer cleaning through a high-pressure nozzle with optical recording of the sewer condition.

A camera is installed at the head of this maintenance and inspection device next to the individual cleaning nozzles that exit to the rear (Fig. 7.4). The camera is either fixed (satellite camera - JT-Elektronik, Lindau) or swivelling (Lindauer bulb - JT-Elektronik, Lindau).

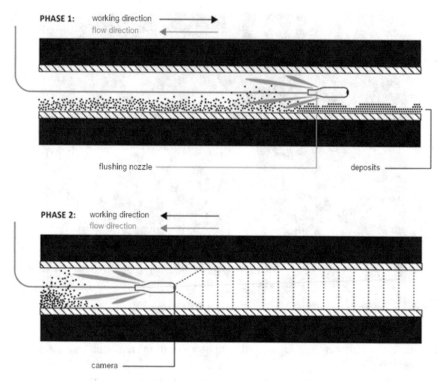

Fig. 7.4 Schematic of High-Pressure cleaning method (FiW)

In Germany, this widely used process for sewer cleaning is called high pressure flushing (HD) process.

After installation in the entry shaft, the nozzle head, together with the hose, is pressed by water jets escaping to the rear against the flow direction from the entry shaft to the target shaft. The water pressure at the cleaning nozzles is usually between 80 and 100 bar. In this phase existing deposits in the sewer are loosened. As soon as the target shaft is reached, the hose is slowly retracted. Due to the water that continues to flow out of the cleaning nozzles, the deposits are also washed away and the driven-on area is cleaned. In this phase the camera inspection is carried out. As a result, the cleaning of the sewer can be inspected as well as the condition of the sewer. A usable camera inspection can be realized at a filling level of up to 20 percent. The satellite cameras from JT-Elektronik are suitable for sewer diameters from DN 70 to DN 300. In addition, there are also cleaning nozzles that can be used up to a diameter of DN 1200. Another argument in favour of this method is that it can be installed easily and quickly with relatively little personnel resources (2 persons). It is not applicable for thicker deposits, as the cleaning performance is not realizable. Furthermore, it is not advisable to use the HD method for already damaged sewers, as there is a high risk that the damage will be increased by the high pressure of the water jet. Care must also be taken in the area of the sewer sleeves so that they are not damaged.

The stresses on the pipe surface are dependent on:

- The water pressure at the cleaning nozzle,
- The amount of water,
- The distance of the cleaning nozzle to the pipe wall,
- The number, the cross-section and the outlet angle of the cleaning nozzles.

China

In China, a sewer inspection cannot be carried out by a complete closure of one or more sewerage sections. As opposed to Germany, the sewer infrastructure is not developed to such an extent that the wastewater can be diverted without problems. This is partly due to the infrastructure itself, and partly to the high level in the sewer network caused by groundwater intrusion. As soon as an area is completely blocked off for inspection purposes, there is a risk of flooding in the areas in front of the blocked-off sewer area within a short time.

At the moment, four different methods are applied in China in the field of sewer inspection:

CCTV-method,

Damage detection by sonar,

GPR-method and.

Periscope procedure.

The most widely used procedure is the CCTV (Closed Circuit Television) method. This procedure corresponds to the sewer robots known here in Germany with integrated rotary head camera. Using this procedure, the crawler equipped with a camera travels over the area to be viewed. Prerequisite for the use of this technique is a low filling level and a prior cleaning of the section to be examined, which cannot be realized optimally in China due to the above mentioned reasons.

Another inspection method used is the detection of damaged areas using sonar. Here a sonar probe is installed onto a floating body which moves through the pipe at high filling levels and inspects the sewer pipe by means of sonar. When evaluating the sonar recordings, it is difficult to distinguish damage to the bottom of the sewer from deposits. There is a great risk of misinterpretation and thus wrong conclusions during sewer inspection.

The GPR method (Ground Penetrating Radar) works in a similar fashion to the sonar inspection method. Only the operating principle of this method is not based on sonar technology, but on the use of radar antennas. Similar to the CCTV method, the GPR method is only feasible at low levels, as too high a water content complicates a radar operation and falsifies the radargrams. Accordingly, the GPR method is only conditionally practicable in China.

Recently, inspection trials have been running in China with a periscope method, which corresponds to the electrical sewer mirroring widely used in Germany. In Germany, however, this method is only used for first sighting and remote sensing. A high quality inspection and exact damage detection, similar to CCTV inspection, is not possible solely with the sewer mirror.

Instructions for Direct Inspection
Germany
With direct inspection, the actual structural condition is assessed by the inspector by means of a local inspection of the sewer system. The inspector enters the sewer system via the manhole and documents the findings and their quantification. A recording in the form of photography and video technology is planned to enable further work steps. According to the German/European standard DIN EN 752:2008–04, direct optical inspection is to be avoided for occupational safety reasons. However, according DIN EN 13,508–01 2013 reads as follows:

> "Should it not be possible to obtain sufficient information through indirect inspections, a direct inspection (e. g. walk-through sewer) may be carried out. Requirements concerning the circumstances allowing for direct inspection shall be taken from national rules or obtained from the competent body".

According to the European standard and the German DWA instruction leaflet, the direct inspection must be carried out by a team of at least two persons. After the sewer has been measured "clear" indicating absence of contamination, i. e. an atmosphere in the sewer which is not dangerous for humans (measurement of gas concentrations such as hydrogen sulphide, carbon dioxide etc.) has been checked, the first inspector enters the section of sewer. This first inspector is equipped with explosion-protected measuring equipment and a portable sewer camera that transmits recordings in real time to the second inspector in the recording vehicle and permits storage of the recorded data. The second inspector evaluates the measurement results in the recording van and gives further instructions to the colleague inspecting the sewer. A radio connection between the two persons is also mandatory. This ensures fast communication and thus work safety, as any emergencies can be communicated quickly [5]. According to the European standard DIN EN 13,508–01 2013, a minimum diameter of the sewer of DN 1200 must be given for direct inspection. Moreover, it must be free of wastewater flow during the inspection.

China
In accordance with the Chinese safety standard for the operation of electrical equipment in potentially explosive atmospheres according to GB 3868–1983, a direct inspection must also be carried out by at least 2 persons. Furthermore, the Chinese CJJ 181–2012 standard specifies four criteria for a direct inspection by an inspector. These are:

- a minimum sewer diameter of 0.8 m,
- a flow velocity of not more than 0.5 m/s,
- a water depth of not more than 0.5 m and
- a filling level of less than 50% as regards the cross section.

In order to detect damages to drainage systems at an early stage, the sewer network operator in Germany usually carries out a condition assessment every 10 years, and even more frequently in ground water protection areas. Furthermore, visual inspections are recommended when acceptance testing of sewer construction and rehabilitation measures are carried out, and also before the warranty period for new construction or rehabilitation measures expires. The results of the recording of conditions are documented in the sewer cadastral. In the case of sewers that have been repeatedly inspected, a recommendation is made to also retain earlier inspection results (Bavarian State Office for the Environment 2010). Due to the status of German standards and the fact that instructions are not legally binding, the aforementioned time intervals of the condition assessment are not considered legally binding either. Therefore, a sewer operator can decide at his own discretion on the frequency of the condition assessment. Basically, a distinction can be made between three procedures. Concerning the fire brigade method, incidents are eliminated with the help of emergency task forces, who are sent to the incidents in order to repair them without regular inspection beforehand. The preventive strategy is characterised by fixed intervals for the inspection, maintenance and rehabilitation of a sewer section. The last and most frequently chosen option is the inspection strategy based on requirements. With this strategy, maintenance and rehabilitation are carried out on the basis of the evaluated findings according to a rehabilitation plan (Pinnekamp 2014).

The Chinese regulation CJJ 181–2012 specifies an interval of five to ten years for an inspection period. However, the interval varies if a structural or functional defect was detected during a previous inspection, which requires a more frequent inspection (up to 1–2 years). Other reasons for reducing the frequency of inspections are:

- sewers placed in specific soils (eroded soil, flowing sand, etc.),
- sewers older than 30 years,
- sewers which are in poor structural condition,
- infrastructurally important sewers and
- sewers with specific requirements.

Due to the legally binding situation of the Chinese regulatory framework, an exact time frame for the condition assessment can be specified here (CJJ 181–2012, 2012).

Instructions for Indirect Inspection
Germany
Indirect inspection is the assessment of the current condition of sewerages by means of an indirect visual inspection, using remote-controlled sewer inspection techniques. In Germany, the instruction sheet DWA-M 149–5 (according to DIN EN 13,508–01 and DIN EN 13,508–02) specifies the field of application for indirect inspections with a nominal diameter for pipes of DN 100 to 1200 (100 - 1200 mm). The equipment of an indirect inspection comprises the following main components: camera system, drive unit, power supply and data transmission as well as the control unit. According to DWA-M 149–5, the camera systems should be equipped with an optical zoom, autofocus and equipment to ensure an upright image. In addition, functions for quantitative condition assessment or position determination are possible, including

localisation sensors and functions for damage measurement, e.g. by laser measurement. Halogen lamps or light emitting diodes are used to illuminate the inspection area. Controllable crawler vehicles can be used as from a nominal sewer diameter of DN 100. Adaptation to larger nominal widths is achieved by mechanically or electrically operated height-adjustable units or different wheel sizes. If the use of mobile camera robots is not possible, there is the possibility of a drive by sliding rods or flushing nozzles. The energy supply of the camera as well as the data transmission are realised via a special cable on a motor-driven cable winch, which is also used for length measurement. The control unit is the interface between camera vehicle and inspector. The inspector can use it to operate the sewer camera and view and record the video material on a screen. The evaluation of the documented recordings is software-supported using a coding system [5].

China

The technical requirements for indirect inspection are described in the Chinese regulations CJJ 181–2012 much like the German regulations. The CCTV contains technically similar main components as those described in the German regulations. It is a mobile crawler equipped with a camera and measuring technology, which captures the surface of the sewer by video technology. The evaluation of the recordings is done by the inspectors in a camera car, who record the condition of the drainage system under consideration of a coding system. The regulations describe a nominal width of at least DN 100 for CCTV inspection. According to the Chinese regulations, the wheels of the crawler must be exchangeable and adjustable in height. Furthermore, the camera should have a lens that can be moved in all directions as well as a zoom function and autofocus. On the monitor of the inspector the date, time and position parameters of the inspection must be made available. A freely adjustable height adjustment of the camera is also required.

Sewer Cleaning
Germany

According to DWA-M-149–5, inspections which fulfil the purpose of assessing the structural condition of an object require a complete and comprehensive cleaning of the object to be inspected in a temporal context to the inspection itself. The temporal connection must be adapted to the operating situation to such an extent that renewed pollution cannot occur. The instruction sheet therefore specifies a lead time for cleaning of 48 h [5]. High-pressure cleaning as well as cleaning procedures adapted to the degree of soiling are mentioned as cleaning procedures. In high-pressure cleaning, a flushing nozzle is moved through the sewer section. This nozzle conveys the impurities and deposits by water pressure in the direction of the shaft, from where they are removed by means of a suction instrument. According to DWA-M-149–5, inspections which fulfil the purpose of assessing the structural condition of an object require a complete and comprehensive cleaning of the object to be inspected in a temporal context to the inspection itself. The temporal connection must be adapted to the operating situation to such an extent that renewed pollution cannot occur. The instruction sheet therefore specifies a lead for cleaning of 48 h

[5]. High-pressure cleaning as well as cleaning procedures adapted to the degree of soiling are mentioned as cleaning procedures.

The intensity of the flushing nozzle should be selected so that the sewer section is free of all detachable pollutions and deposits and a full inspection of the sewer wall is possible. The intensity results from the number and size of the flushing nozzles mounted in the nozzle head, which can be optionally mounted in such a way that they can bring a lot of intensity and cleaning power for all areas of the sewer walls. A lower intensity of the flushing nozzle is required, for example, in sewer systems where damage is known or suspected due to obvious circumstances [5]. For an optimal investigation and assessment of a sewerage, attention should be paid to well-trained and experienced cleaning personnel (Vogel 2007). According to DWA-M 149–5, suitable cleaning instruments should be available, even during the inspection, in order to be able to clean the sewer if necessary [5].

China
The basic principle of sewer cleaning carried out in China is similar to that of German sewer cleaning. In China, this cleaning step is also preceded by the actual inspection in order to be able to better assess the condition of the sewer. Similar to Germany, there are also different cleaning nozzles in China, which can also be varied in cleaning performance by changing the nozzles. Unlike in Germany however the loose deposits are not sucked off at the next manhole, but are dragged along further. In addition, the flushing vehicles and the installed pumps cannot be compared with those used in Germany. There is a clear difference here, if only in terms of the technical equipment and the cleaning performance to be achieved.

A further problem in China is the lack of staff training. Usually the staff is not aware of the sense and purpose of sewer cleaning and therefore there is a lack of awareness for high quality work.

7.2.3.3 Coding and Classification of Condition Recording

An important tool for managing data of drainage systems is the classification and coding of damages and structural elements. Coding is a system that reduces and standardises the complex information of sewer condition recording to a few pieces of data. This enables expert persons to gain a quick and objective insight into the condition assessment.

Germany
In Germany, DIN EN 13,508–02 and DWA-M 149–2 deal with, among other things, the standardised coding of the sewer condition assessment. First of all, the main code describes the findings made during the status recording. The main codes are divided into those for structure, operation, inventory and other categories. Several main codes can be applied to one element in the drainage system. As an example, a lateral house connection in sewerage section is marked with a structural main code "BCA" and its faulty connection entering the sewer with a code for evaluation of the

Table 7.1 Example for coding in Germany (cf. [3])

Longitudinal position	Line damage code	Main code	Characterization		Quantification		Location on the perimeter		Connection	Photo reference	Video reference	Annotations
			1	2	1	2	1	2				
16,5		BCA	E	A	100		9				00:12:20	
16,5		BAG			50		9				00:12:20	

situation "BAG". Both main codes thus refer to the same event but describe different findings. (The exact origin of the main code can be found in DIN EN 13,508–02 and DWA-M 149–2).

Further coding is added to characterise the finding. The characterisation consists of none, one or at most two details which explain the finding in detail. In the present case seen in Table 7.1, the characterisation 1 with the letter E describes the type of connection. Therefore this is a structural characterisation of the finding. The second characterisation, marked with the letter A, gives the additional information that a blocked connection is involved. As a further coding, details quantifying the finding are added. Quantification for structural recordings includes, for example, the diameter of connections and for stocktaking surveys, for example, the reduction of the cross section of the sewer. Further information in the coding system are positional data. To determine the location of the detection, these are specified in metres [m] in the longitudinal direction of the distance from the reference point of the inspection. DIN EN 13,508–02 specifies various procedures for selecting the reference point. The most frequently selected reference point is the inner wall of the initial node, such as shaft, inspection opening, outlet, etc., in which the position of the sewer to be inspected is integrated. As a further description of the position of the finding, the positioning on the section wall, using the dial reference (time), is determined. This in turn is determined by the angle, which results from the centre of the cross section between the finding and the sewer vertex [3].

China

In China, the CJJ-181 standard is concerned with the uniform coding of channel status recording. If the standard is complied with, the basic information is first of all uniformly recorded as master data. This data can be adopted from the system and adapted if the sewer database is properly managed. Basic information on the examined sewerage system (number, age, material, nominal width, length, initial depth, final depth), the selected inspection procedure as well as its realisation (date, time, inspection direction) and the operators involved in a sewer inspection (name of the inspector, the supervisor and the inspector) are given.

In the CJJ-181 standard, a damage code describes the finding that was made during the condition assessment. The damage codes are divided into structural and functional damage parameters. Structural damage is damage to the structure of the sewer such as breakages, cracks, fissures or deformations. Functional damages primarily influence the function of the sewerage through external factors such as deposits or root ingrowing. After a definition of the damage, the damage is classified into 4 categories. The categorisation is based on quantifiable arguments. Each damage category is assigned points per damage, which serve the subsequent condition assessment of the drainage system (Table 7.2).

In addition to the damage codes, the Chinese standard specifies codes for the recording of so-called operational status information, which describe the start and end position of the damage or which give an abort of the inspection with a corresponding reason (e.g. camera under water). The code of a damage to the pipe system is assigned to the start position if it is longer than 1 m and to the end position if the length of the damage is less than 1 m. There are also codes for special constructions and secondary installations such as inverted siphons or shafts hidden on the surface.

Table 7.2 Examples to describe the structural and functional damage of a drainage system according to the Chinese standard CJJ-181

Type of Damage (structural)	Damage Code	Definition	Category of Damage	Description of Categories	Points of Damage
变形 Bian Xing Deformation	BX	The sewer is deformed by external force	1	The deformation is less than 5 % of the pipe diameter.	1
			2	The deformation is 5 to 15 % of the pipe diameter.	2
			3	The deformation is 15 to 25 % of the pipe diameter.	5
			4	The deformation is more than 25 % of the pipe diameter.	10
Type of Damage (functional)	Damage Code	Definition	Category of Damage	Description of Categories	Points of Damage
沉积 Chen Ji Deposits	CJ	Deposits on the invert of the sewer	1	Deposits in the height of 20 -30% of the sewer	0,5

7.2.3.4   Conclusion

The most obvious difference between the German and Chinese regulations is a different handling/ viewing of direct inspection. For safety reasons, the German standards and instructions describe how to avoid direct inspection, which is not the case in the Chinese regulations under consideration. Furthermore, both sets of regulations specify a minimum diameter as a criterion for feasibility. Here the German minimum value is 50% higher than the Chinese one. Coupled with the other criteria regarding water level and flow velocity on the part of the Chinese plant, conditions arise under which the safety of the inspector carrying out the work cannot be guaranteed. Due to the preceding cleaning with regards to the German inspection, as well as the watertight sealing of the pipe work against further inflows, the safety conditions for the inspector in Germany can be considered to be safer. However, occupational health and safety is important in both variants, as can be seen from the need of protection through teamwork.

In both sets of regulations under consideration, the operating positioning of the camera is the same for indirect inspection. Accordingly, the lens must be centred in the direction of inspection and in the centre of the sewer cross section, therefore in the centre of the profile's cross section. The DWA-M 149–5 first specifies an approximation to the location of the finding in axial direction by swivelling the camera during a sighting. Then the camera should first be swivelled in a horizontal position and finally the lens should be centred on the object in a vertical direction. For this purpose, the regulations state the reason for the orientation aid for the inspector or the subsequent second observer (engineer). Such a regulation is not mentioned in the Chinese reference works. The CJJ 181–2012 describes a different starting situation for the indirect inspection measure than is the case in German. Accordingly, it is not planned to clean and block further inflows in the course of the standardised indirect inspection. Therefore, the indirect inspection, just like the direct inspection, is carried out during operational mode, provided that a certain criterion is fulfilled. This criterion is described with a maximum filling degree of the sewer cross-sectional height of 20%, with respect to the cross section. If the permitted filling level is exceeded due to normal operation, suitable measures must be taken to reduce the flow height down to the specified criterion. In CJJ 181–2012, several reasons for a stop due to bad conditions are listed. A stop of the indirect inspection is required if an obstacle prevents the camera from moving forward, if the lens of the camera is dirty or overflowing with water, or if visibility is impaired by the appearance of fog. The speeds at which the sewer robot moves in the sewer section is regulated to a maximum speed of maximum 0.1 m/s for a diameter $\leq$ DN 200, as well as not more than 0.15 m/s for diameters > DN 200 with the direction of movement in the direction of flow (CJJ 181–2012 2012). The HKCCEC2009 also mentions a third speed in addition to the above-mentioned speeds, for an application in the DN > 300 range. There the maximum speed is limited to 0.2 m/s. For comparison, the German version has omitted a recommended speed specification. The only speed mentioned is one that allows the inspection to be followed without interruption on the basis of the recordings. This means that a moderate speed is chosen. This allows a secondary evaluation based on the given conditions, such as the image rate of the camera and visibility conditions [3].

7.2.3.5   Management / Financing

Sewer Database and Sewer Information System
The main task of a sewer database (SDB) is the documentation and updating of existing data and inspection results to ensure a coordinated and future-oriented maintenance of sewerages. The SDB provides in alphanumeric form the basis for the location verification, planning, construction, operation and maintenance of the drainage system. Consequently, it provides the basic prerequisite for maintenance and plant monitoring. Additional software for hydraulic network calculations, valuation or condition assessment is linked to an existing SDB and serves as an instrument for the engineering processing of the inspection results. In order to check the data and to process them further, today's technical possibilities require a graphic representation by linking to a sewer register [Pecher 91a]. The sewer register is connected to existing sewer databases by means of an appropriate interface configuration. The sewer register offers advantages when applications with large amounts of data and short access times are required. It offers a clear data management with high user comfort [ATVA145]. The sewer register offers the possibility of linking data, texts and images with their geographical location to one information unit. Together with the indirect discharge cadastre, which is used to monitor wastewater collection from industrial companies, it is also referred to as a sewer information system (SIS) [Sawat94] [FI-DWInf].

A sewer information system is thus a system for the recording, keeping, maintenance, illustration, analysis, processing and exchange of information on drainage systems outside of buildings. The quality of the results produced by a SIS is derived from the quality of the geodata stored in the sewer database. As these data serve, for example, the documentation of inventory, are used to produce maps or provide information for statements and expert opinions, as well as being the basis for planning and model calculations, they represent an enormous value. Maintaining these values and ensuring the quality of the geodata in the procurement, creation and administration of the data are part of the core tasks of every SIS and are the prerequisite for the efficient use of a sewer information system. Sewer information systems enable, based on a central sewer database, the creation of comprehensive and over-arching inventory documentation and operational information systems through software for sewer inspection and for the processing and evaluation of inspection data.

Software for Sewer Inspection and for Processing and Evaluating of Inspection Data
The aim of the sewer inspection is to derive the documentation of the structural condition from the recommendations for action to maintain the sewer system. A uniform, generally valid and automated description of the different conditions in the sewer

system is mandatory. The coding system to be used is described in DIN EN 13,508–2, in conjunction with the DWA leaflet DWA-M 149–2, which provides application recommendations (see chapter 2.3.1). The description of the condition is done by means of a main code, characterisation and a corresponding quantification. Operators can learn sufficient expertise and the application of the code in special "Sewer Inspection Courses". The digital sewer register is maintained in a sewer information system. In addition to the pure plan information, such systems provide the data basis for the manifold tasks of sewer operation. The hereby listed attributes concerning the objects of the sewer system serve as a basis for the sewer inspection. The sewers and structures to be inspected are digitally transferred to the sewer inspector. As described in leaflet DWA M-194–5, efficient and economical data management and a continuous flow of information in digital form is required. This data cycle from the SIS to the inspection vehicle and back again is outlined in the DWA-M 150 information sheet. Here, the master data of the sewer sections and structures are transferred to the inspection vehicle in digital form. The operator validates the master data on site and extends the data by the description of the current condition. At the end of the sewer inspection, the data are digitally transferred again for further processing.

On the inspection vehicles or mobile camera systems, therefore, a database-supported software must be used to enable digital data exchange and electronically supported damage description. Inspection software products are highly specialised products which, on the one hand, have to communicate with the hardware of the inspection vehicle/ crawler and, on the other hand, have to take into account the common specifications for the coding system, data management and data cycle. Inspection software products are developed or distributed by several hardware manufacturers. There are also free software manufacturers who have adapted their inspection software to the hardware of different manufacturers. An inspection software must document the condition of the sewer system. In addition to the coding, a video and optionally a photo is usually added to the status. To be able to locate the documented status, the station of the condition must also be specified. Through the coupling to the BUS system, the inspection software receives relevant information, such as the distance travelled, or sends information for the insertion of text into the video. A further important task is the control of the video recording. For this purpose, a video grabber must be available to encode the analogue video signal into a file format. The video grabber provides an API to control the video recording and the quality of the video file.

Cost Implications of a Regulated Recording and Management System for Sewer Data

The construction and operation of the drainage facilities, as well as their maintenance, require a large amount of financial resources. At the same time, the public infrastructure of drainage facilities is often one of the largest asset of cities and municipalities [8]. Therefore, the economic aspect plays an important role for the operator. It is important to keep the costs as low as possible in order to ensure efficient use of the financial resources of the cities and municipalities. It is also in the interest of the cities and municipalities to maintain the financial value of the sewerage system for as long as possible. When the sewerage system is built, it is given a predicted operating life, which determines the duration of depreciation. It is therefore the intention of the operator to maintain the condition of the sewerage system as good as possible in order to extend the depreciation period, if necessary, and thus reduce its annual costs, or to ensure that the sewerage system continues to exist beyond the depreciation period. Further costs are incurred by the municipalities and local authorities as a result of the rehabilitation or even new construction of damaged drainage components. Since it is logistically and economically impossible to keep all sewers in a faultless condition at all times, plans must be drawn up for the necessity of rehabilitation measures. These plans provide information about the urgency of possible measures and can be used for a direct comparison between sewer sections in order to decide which sewer section is most in need of rehabilitation. The basis for these plans is a detailed and precise condition assessment, followed by a technical classification and subsequent evaluation of the conditions.

In order to avoid a conflict of interests, a differentiation between the persons carrying out the condition recording and the classification, i.e. an evaluation of the condition is recommended (LfU 2010; ATV-M 149, 1999).

During the operation of the sewer system, questions of economic efficiency further arise. For example, the operation of wastewater treatment plants, whose efficiency is directly dependent on the throughput. Extraneous water entering the sewerage, which for the most part does not require treatment, is also treated in the treatment plants and occupies capacity. These additional capacities result in increased costs for operation. In order to keep the costs for the transport and treatment of the wastewater low, the condition of the sewerage must be checked in order to detect damaged sewers and reduce the infiltration of extraneous water to a minimum. It is also important for the operation to maintain the structural conditions in order to enable proper operation under favourable hydraulic conditions and to ensure that parameters such as flow velocity, shear stress and gradient are within a desired range. This is guaranteed by a condition assessment or a rehabilitation in the later course of the project.

7.3 Application of Advanced Sewer Inspection Technologies in Jiaxing

7.3.1 General Notes

Analysis of the Conditions on Site
The initial meeting in Jiaxing, China was held to investigate the situation on site.

The examination showed that the present sewer systems lack a proper drainage of upcoming sewage water and entering groundwater. Hence, a certain water mark is always present in the system that allows solid matters to be collected in the sewers and shafts. This situation complicates the inspection and excludes several inspection systems established in Germany.

The meeting indicated that the FastPicture method is already available and is being used by the Chinese partners. Even though the system provided by JT-elektronik offers higher video quality as well as robustness, this option was rejected.

It was decided to design a new and customised system based on the SKI method. To meet the requirements on site properly, the new system should comprise the following specifications:

– possible visual inspection of the sewer above water level
– adjustable camera system in height depending to the water level
– adjustable transport system to the sewer dimension
– portability of the inspection system
– simultaneous cleaning and inspecting of the sewer

In the following the design and implementation of the first prototype as well as the commissioning and tests are outlined.

Application of prototype I shows the former prototype of the inspection system. Now the transport vehicle is based on a sledge system with three blades (see Fig. 7.5). The blades facilitate the movement of the system in the sewer by slicing the solid deposits. Additional wheels attached to the blades serve to minimize friction and therefore abrasion of the sewer material. The size of the system is continuously variable corresponding to the dimension of the sewer.

In total, two setups were built to cover the sewer diameters from 150 to 300 mm and 400 to 800 mm, respectively. Furthermore, the symmetric arrangement of the three blades offer an unproblematic removal of the system in case of an accident (e.g. tip over of the inspection system).

The drive of the system is based on a customised cylindrical nozzle (see 7–10) that offers the combination of controllable drive and cleaning by adjusting water flow and pressure. The insets of the nozzle are available in VA-steel and ceramic which is more robust against abrasion by solid particles in water. The angles of the inlets were adjusted to obtain an optimised jet for driving (wide angle) and cleaning

Fig. 7.5 Prototype I (FiW)

(small angle). The nozzle can be driven by a ¾" or 1" hose, depending on the scope of application. To prevent tilting of the inspection system, caused by the hose, a rotating joint was integrated (Fig. 7.6).

Visual inspection is conducted by use of a pan and tilt camera mounted on the transport system. The height of the camera can be adjusted to be centrally oriented in the sewer with a diameter range of 150 to 800 mm. This ensures an unobstructed view independent of the water level. The camera offers a video signal with a resolution of 500 × 576 px, manual focus and automatic aperture manipulation.

The system provides an integrated cleaning function during inspection. The water flow through the nozzles induces a jet removing water and solids from the camera surface and lens when swivelling the camera head by 90° aside. This ensures a clear visibility throughout the inspection.

The system can be controlled (i.e. light, pan, tilt) by a portable structure (see Fig. 7.7, left) with an integrated battery pack, a tablet for video inspection, and

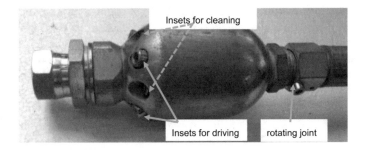

Fig. 7.6 Customised nozzle for prototype I. (JT Elektronik)

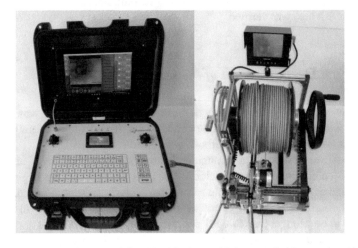

Fig. 7.7 Portable control system (left) and cable drum with integrated video monitor (right). (JT Elektronik)

Fig. 7.8 Test of prototype-I
in Wiesbaden. (JT
Elektronik)

a human–machine-interface for camera control via buttons, joysticks and a touch panel. The connection to the camera is realized by a cable drum (see Fig. 7.7, right) with a cable length of 150 m, a video monitor, and an integrated cable length counter being displayed in the control system.

Commissioning and testing of the system was executed under ideal conditions on the company premises of JT-elektronik GmbH Lindau and on a test area in Wiesbaden, Germany (see Fig. 7.8). The vehicle was placed into sewers with different diameters ranging from 200 to 900 m. The sewers partly contained water and solids up to 15% of the diameter. The system was connected by a ¾" water hose to a flushing vehicle. The vehicle was equipped by JT-elektronik GmbH with a high-pressure pump and tank system providing a water flow of 120 l/m and maximum water pressure of 160 bar. The system was successfully commissioned and the functionality (i.e. proper driving) was verified.

7.3.2 First Demonstration of the Inspection Process

To test the prototype under real conditions, a meeting in China was arranged (see Fig. 7.9). The participants comprised Tobias Jöckel (JT-elektronik), David Eisenhauer (bluemetric), Jan Echterhoff and Marc Jansen (FiW).

The tests were executed in a municipal sewer system in Jiaxing, China which reflects the conditions in Chinese cities. The tests showed that the huge depth of the shaft and the absence of climbing irons, particularly present in China, severely impede the insertion of the inspection device. Furthermore, the soiling in the sewer was found to be stronger than initially assumed. During the tests it was apparent that the height offsets throughout the sewer as the major challenge, since the inspection system could get stuck. To avoid this and improve the handling of the inspection system the vehicle was optimised in further steps described in the following.

Fig. 7.9 First tests in Jiaxing, China. (FiW)

7.3.3 First Assessment of the Inspection Process

The existing water level and the height offsets in the sewers require a floatable system. Therefore, prototype I was extended by two floating bodies, see Fig. 7.10. Floating bodies with increased length serve to better stabilise the system during floating; however, the increased length renders handling more difficult when the system is inserted into the sewer, especially when having small inlet diameters. Therefore, the system provides a mounting for two different sizes of floating bodies. The first one is applicable for sewer diameters from 200 to 400 mm and the second one for diameters ranging from 500 to 1500 mm. The floating bodies can be mechanically adjusted to the diameter of the sewer. A keel for stabilization during floating was attached.

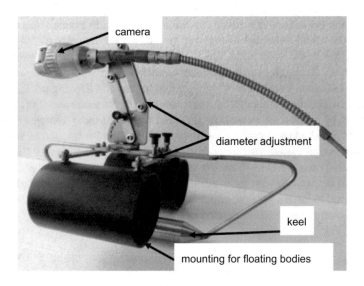

Fig. 7.10 Inspection system prototype II. (JT Elektronik)

Fig. 7.11 Prototype II including floating bodies and nozzles. (JT Elektronik)

Due to the heavy soiling in the Chinese sewer systems the nozzle was replaced by a combination of two coordinated nozzles, see Fig. 7.11. The first is located in proximity to the vehicle right below the water surface and serves for driving the system. It also ensures a stable floating of the system itself. The second nozzle is attached to a tube approximately 50 cm behind the vehicle. This nozzle is pulled down to the bottom of the sewer and is therefore used to remove the soiling and clean the sewer. The system is also equipped with a rotating joint (comparable to prototype I) to avoid tilting of the inspection system.

To test the system under realistic conditions a test field in Munich, Germany was used. The test field consisted of a sewer system with a diameter of 600 mm and contained gravel and water up to 30%. The tests were executed in cooperation with Huber Linden, Reinigung und Entsorgung GmbH.

Figure 7.12 shows the test setup with the inspection and flushing vehicle to drive prototype II. The inspection system was inserted into the sewer (see Fig. 7.13) and was driven by a constant water flow provided by the flushing vehicle. The video signal (see Fig. 7.14) was captured with the equipment in the inspection vehicle.

The inspection system was successfully tested and the functionality (i.e. driving and cleaning) was verified. The resulting speed and the cleaning performance appeared to be sufficient for the conditions in China. The optimised nozzle combination proved to have a positive effect on the stability of the system while floating. The video quality and the illumination of the sewer were satisfactory.

Fig. 7.12 Test setup in Munich. (JT Elektronik)

Fig. 7.13 Inserting the prototype II into the sewer. (JT Elektronik)

Fig. 7.14 Video signal of the inspection. (JT Elektronik)

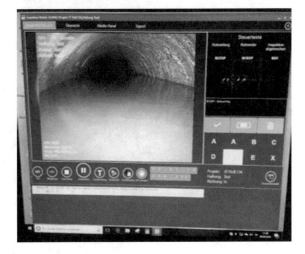

7.3.4 Second Demonstration of the Inspection Process

Due to the positive results of the previous tests, prototype II was transferred to China for final assessment.

The system was presented to the Chinese partners and a training was held to properly operate the system (see Fig. 7.15 and Fig. 7.16).

Final tests took place in the municipal sewer system. The insertion of the inspection system was practicable due to the high variability of this system in width and height. Various tests in canal systems with different diameters were carried out in cooperation with the Chinese partners. The handling and the evaluation of the actual performance of the system was practiced intensively.

Fig. 7.15 Final project meeting in China. (FiW)

Fig. 7.16 Training of Chinese partners in operating the inspection system. (FiW)

7.3.5   Second Assessment of the Inspection Process

The tests show that mechanical optimisations of the system described above improve the handling of the device significantly. Even when solid particles are given, the floating bodies ensure the movement in the sewer with a water level up to 60%. The cleaning and driving performance of the nozzles were proven to be sufficient.

The applicability of the system to sewers with different diameters enables a flexible application to a wide range of sewer systems in China. Additionally, the mechanical structure of the systems allows a safe recovery in case of damage. The camera system enables a visual inspection of the sewer during cleaning, which is time-saving.

Beside the various advantages the following limitations were recorded.

Due to the overload of the Chinese sewer systems it is necessary to clean the sewer before inspection. Since there are no vacuum trucks available in China to remove the solid material from the sewer, cleaning options are limited.

The presented inspection system can be used for the majority of sewer systems in China. Nevertheless, the water and dirt levels vary within different sites which makes the system a system of limited application (e.g. low water level impedes floating of the device). Therefore, a combination of prototype I and II is advised. Subsequent to the tests, the system was handed over to the Chinese partners to proceed with further investigations.

7.4   Management Systems for Sewer Data Evaluation

7.4.1   Technology Development

Requirements
Within the framework of technology development, the choice of hardware and software depends on several factors. There are on the one hand project specifications, which were discussed in advance by the project partners involved and adjusted to the current situation on site. In this context, local conditions such as for example the level of knowledge of the skilled workers play a major role. On the other hand, a tailor-made condition abbreviation system has to be developed for this project, which is adapted to the local requirements.

Prior to the start of the project, the local conditions in Jiaxing were studied and the requirements for a sewer inspection system were defined through several site visits and meetings with the project participants. Thus the skilled workers should be able to independently carry out inspections under the given conditions with the developed system. Instructions and the daily work should be possible to be passed on to each other without the support of skilled workers. The hardware should be compatible with the local infrastructures, e.g. flushing vehicles, or be connectable to them. Furthermore, it should be possible to build up a permanent data stock through data exchange, from which the condition of the local sewer system can be derived.

From these requirements, framework conditions for the implementation of the overall system can be derived, e.g. which hardware and software components should be used.

Local Boundary Conditions

Apart from the project specifications, the local conditions must also be taken into account. As opposed to Germany, training and a certain level of qualification of the skilled workers cannot be assumed, so that the level of knowledge of the skilled workers varies greatly and the handling of the hardware and software should be kept as simple as possible. In addition, cultural and administrative differences in the field of sewer inspection in China lead to different approaches and operating sequences. For the test runs during the first practical test in Jiaxing, the condition was communicated in advance that the sewer sections to be tested should be pushed off. Since the administrative classification of the responsible wastewater companies is not done locally, but according to the diameter of the sewer section, the sewer sections are still in operation when the test runs are started. This necessitates an adaptation of the system to the above-mentioned conditions.

Condition Abbreviation System
Hardware

In addition to the specific JT Elektronik hardware (slide construction and case system), the Tablet Pokini Tab A10 with the following specifications relevant for the technology development was used:

- Display size: 10.1" with 1920 × 1200 pixels
- Operating system: Windows 10

The size and resolution of the display determines the available space for the software application. For this reason, the UI design of the software was based on a minimum resolution of 1920 × 1200 pixels and, thus, a complete display cannot be guaranteed at lower resolutions.

The decision to use Windows 10 as operating system is based on the wide distribution, the regular supply of updates as well as the positive experiences with previous software projects for sewer condition recording of bluemetric software GmbH. Since the Chinese project partner is already using the operating system in other areas, it can be assumed that they have knowledge in handling of the operating system.

The used camera delivers an analogue video signal, which must be converted into a digital signal for further processing. An external analogue-to-digital converter (A/D converter) is used for this purpose. In agreement with JT-Elektronik, an A/D converter from Sensoray (Model 2253) was employed here. This model has a low latency of the video image, so there is almost no delay between the real video image and the display in the software. This is especially worth mentioning in connection with hardware control, since a time delay of the video image does not reflect the current orientation and position of the camera.

In addition to the transmission of the video image, the software must receive the current and already travelled track of the hardware for complete documentation of the sewer condition. This is the only way to locate the relevant position in the sewer during later evaluation. The connection was made via a serial BUS system developed by JT-Elektronik for transmission of the track length. Should the hardware detect a change regarding the length of the track, this information is transmitted to the software via the BUS data connection and thus displayed by it.

Software

As previously explained, a tablet with the Windows 10 operating system was selected. The possible software technologies for creating software are generally depending on the operating system used. bluemetric software GmbH has decided to develop software based on .NET technology. .NET is a platform developed by Microsoft for the development and execution of application programmes and is integrated in the Windows operating system. The advantages of software developed on the .NET platform are the optimal integration into the operating system, already existing functionalities that do not have to be implemented by the user, independence of the Windows version used as well as continuous further development and error corrections by the manufacturer with automatic updates via Windows updates. Windows Presentation Foundation (WPF) was used for the development of the software interfaces. WPF is included in the .NET platform and allows free and individual design of the individual interface elements such as buttons and selection lists. WPF furthermore allows the separation of interface (View) and logic (Viewmodel). The interface can thus be edited independently of the logic. This was strictly adhered to during the development of the software. The relational database system SQLite was used for permanent storage of the data. SQLite has the great advantage as compared to other database systems that it does not require any additional server software. For use, only a freely available programme library must be integrated into the executing software, which then provides all necessary database functions. In addition to the freely available SQLite programme library, the following non-free programme libraries were used: Syncfusion and Datastead TVideoGrabber. While the Syncfusion programme libraries provide a variety of features that extend the .NET platform. Syncfusion functions were used for developing the software to generate PDF documents because the .NET platform does not offer them. In contrast, the digital video signal must be stored in a video file to fully document the condition of the sewer. To reduce the file size of the video file, it must be compressed. Video codecs are used for compression. Since the .NET platform does not provide any functions for this, the programme library TVideoGrabber from Datastead was used.

Object Structure

All collected data is organised into projects by the developed software. The set-up of a project is always the starting point of every sewer inspection. The projects can be organised here according to any thematic scheme, such as for example streets, villages and towns or by date. As master data, they themselves contain simple attributes such as name, creation and modification date. Figure 7.17 shows the structure of projects and contained objects.

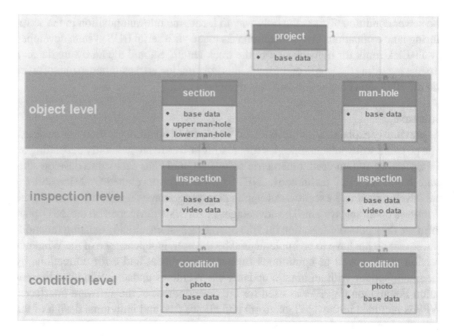

Fig. 7.17 Project structure

In each project, any number of main sewer objects can be recorded in the object level. According to the project specifications, only sewer section and manhole objects are implemented here for the time being, since in China inspections are mainly carried out in the public main sewer system and not in the house connecting pipes. Each sewer section contains master data based on the DWA and links a manhole at the top and a manhole at the bottom. Standardised master data can also be added to a manhole. On the inspection level, both object types can include any number of inspections. The inspections of both types of objects also have DWA standardized master data and a linked video for condition recording. The last level is the condition level, on which there can be any number of conditions per inspection. The conditions contain - similar to the condition objects of German standards - typical master data such as stationing and condition abbreviations. As a further possibility a photo can be linked. A table with the master data of the conditions can be found in Annex B, the structure of the condition abbreviations is explained in chapter 4.1.1.3, whereby the conditions contain condition abbreviations according to the table in Annex D.

Process Programme Sequence

The programme (see Fig. 7.18) consists of a main view in the middle, tab elements in a bar on the left, a title bar in the upper area and an object list in the right area. The tab bar allows the user to change the view at any time. In this way, for example, settings - such as the language - can be made during an inspection or the master data of an object can be retrieved or adapted.

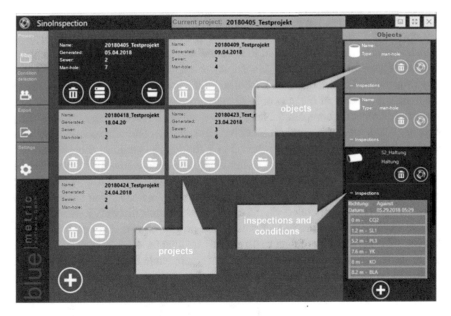

Fig. 7.18 Programme surface/interface - project view

The programme sequence for carrying out inspections is shown in Fig. 7.19. The starting point is always a project.

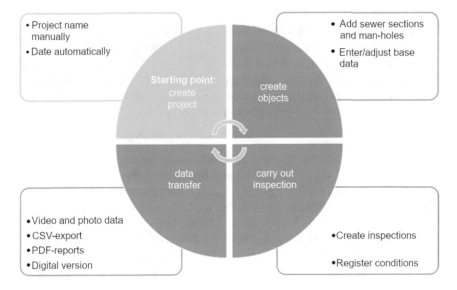

Fig. 7.19 Programme sequence/to carry out inspections

In the project view, the user can start a new project using the button in the lower left corner. As soon as the user starts a new project or opens an existing project, the view is automatically switched to the condition recording view and the existing objects of the current project are loaded. After opening the project, in the next step new objects can be generated via the button in the lower area of the object list. Then the master data view of the new object opens. After confirming the master data, the set-up of an object is completed. Next, inspections can be carried out for the existing objects. When an inspection is started, the video in the condition recording view is started automatically. Now, while the video is running, a condition abbreviation can be selected in the right area, next to the video, and thus a condition status can be added. The video window is reduced in size and the master data of a condition can be entered. After performing sewer section inspections, a sewer section report with all relevant data can be generated via the export view. Additionally CSV files (Comma Separated Values) with the master data of the project objects can be given. These should offer a possibility of data transfer (see also chapter 4.1.4.5 Data Transfer) for the customer, also because these files are easily adaptable from outside. In addition to these two exports, projects can be exported as a whole, with the sewer section reports and CSV files also included.

UI-Concept
The development of the user interface (UI) plays a special role in the development of the software. The UI is the programme interface on which the interaction between user and machine (human–computer interaction) takes place. The goal is an intuitive and efficient operation of the software with minimal need of instruction or training.

Touch-Control
The tablet used has a touch display and thus allows operation with the finger instead of the mouse pointer of an externally connected mouse. When the screen is touched, a click action is triggered for the UI element in this position, similar to the mouse click when operating with the mouse. However, with touch operation, the precision is significantly lower as compared to mouse pointer use. As a result, the operation of small, closely spaced UI elements with the finger is more error-prone.

Size of Control Elements
Controls are all UI elements that interact with the user, including buttons, selection lists, lists, text fields, etc. In the case of touch operation of the software, the control elements must be designed and placed in such a way that error-free operation is ensured. The buttons used have a diameter of approximately 1.5 cm and a minimum distance of approximately 0.5 cm. With the selected diameter, the buttons are not too small and can easily be hit with the finger, the distance avoids accidentally hitting an adjacent button. When developing the UI, overlapping programme windows and views were avoided as far as possible. There should be only one UI layer in which the current view is shown.

Integrated Programme Navigation

The software provides the user with programme guidance, which leads to a clear and intuitive programme flow and minimizes incorrect entries. In the main view the current topic to be worked on is displayed, this must first be completed before the next processing step can be carried out.

Simplified Condition Abbreviation System

Chapter 2.3.3 described the structure of the Chinese condition abbreviation system. In the software, the main code must be selected in the first step; if it is an error code, the level of deficiency must then be selected (see Fig. 7.20). If the main code is an event, no further specification is necessary (see Fig. 7.21). Finally, the selection must be confirmed.

Localisation

The development of a localisation concept is decisive in the context of the research project, as the software to be used must be understood by Chinese specialists and decision-makers, as well as by German project partners who instruct the Chinese specialists on site. This is also necessary in view of further international software projects, where the possibility should exist to make other languages available. In the software, it is therefore necessary that the language can be changed during runtime and that all texts are changed to the selected language. For this purpose, a selection menu is available in the settings view which can be used to specify the desired language. By changing the language, there must be dynamic text elements in the software views that display the text of the selected language package. The concept of DataBinding is used for this purpose, in which adjustable variables are used in the views, which are controlled by the logic in the background. The language texts displayed on the user surface are organized in XAML-based language files, in which the individual language text elements are each stored in a key-value pair. Here the key is embedded in the views and the associated value from the language file is then displayed for the user. This procedure/ process also offers the possibility to adjust language texts if they contain errors or if the text contents of a text element have to be changed. A further advantage is that the texts can be adjusted on site. No special software is required for adjustments, changes can be made with any text editor which is available on all operating systems. Furthermore, this concept allows the easy addition of further language packages by generating language files with the same keys and values adapted to the new language. This procedure also has the advantage that Chinese characters, i.e. texts in Unicode16 format, can be inserted

Fig. 7.20 Level of deficiency

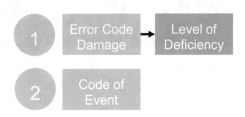

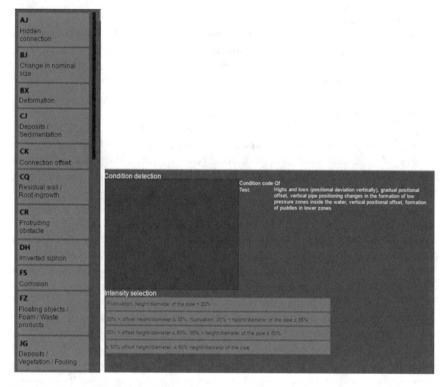

Fig. 7.21 Condition code specification

directly into the language files, which are subsequently only displayed by the software in their original form. Figure 7.22 shows the comparison of the condition recording view with concerning German and Chinese language settings respectively. Further programme views with localisation comparison can be found in Appendix E.

Fig. 7.22 Programme view in German and Chinese language

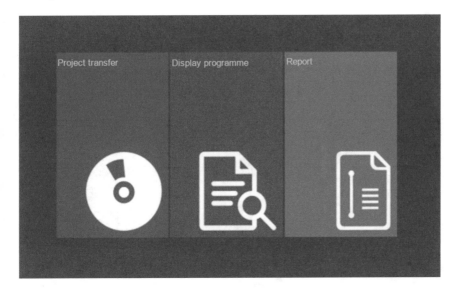

Fig. 7.23 Interface of data transmission

Data Transmission
In order that the recorded data can be used for further processing, the software offers various possibilities for data transmission as shown in Fig. 7.23.

Project Transfer
During project transfer, all data recorded is displayed. As opposed to the exchange formats used in Germany based on DIN EN 13,508–2, there is no guideline in China for a standard based exchange format. However, in order to be able to pass on the recorded data, the export of a text file in CSV file format was implemented. All master data and sewer condition data recorded with the software are displayed in this file. The file can already be opened and read with simple text editors. Many programmes allow the import of files in CSV format.

Display Programme
The visualisation programme allows the data to be transferred to an independent programme for viewing the recorded data, yet the data cannot be changed. The display programme (see Fig. 7.24) can be run independently from the software for recording the sewer condition and thus allows flexible transfer to other computers without the additional, prior installation of further programme files.

Reports
With the software, a report can be created in Microsoft Word and PDF file format for each inspection performed. The report contains all relevant documented data of the created objects. The recorded condition codes are displayed in both tabular and graphical form (see Fig. 7.25).

Fig. 7.24 Display programme

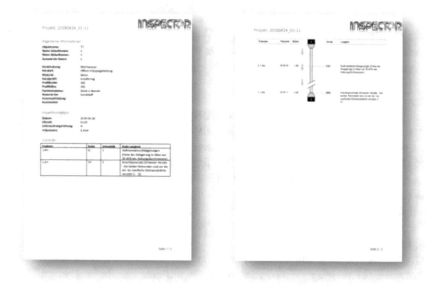

Fig. 7.25 Documented condition codes in tabular (left) and graphical form (right)

7.4.2 First Demonstration of the Inspection Procedure

In the scope of the first practical test in October 2017, bluemetric made a presentation for the project partners in Jiaxing, which served to illustrate the programme structure and the backgrounds. The experience gained with the standards-based condition recording in Germany was used to establish a connection to the set-up of the software for the practical test in China. Accordingly, the content was the programme flow of the software developed for this project and the structure of the condition codes used on the basis of the DWA, since the project partners initially had no specifications in this regard. In addition, the touch-based operating concept of the program was explained in detail so that the project partners could gain an insight into the handling of the software.

In the course of the first practical test in Jiaxing, the complete system was set up after unpacking the hardware and software. The commissioning was then carried out together with the Chinese skilled workers to explain the procedures of a condition recording with the software and hardware. The hardware components of JT elektronik GmbH were connected to the Chinese flushing vehicle and software adjustments were made to correctly connect the hardware.

Subsequently, a test run with the inspection sledge vehicle of JT elektronik GmbH took place in front of the premises of the Chinese partner company.

Following the practical tests, the findings were conclusively exchanged and feedback was given by the project partners on useful adaptions for further practical use.

7.4.3 First Assessment of Process

After completion of the first practical test in China, the on-site findings were analysed and the further procedure was determined. It was determined that due to the practical conditions on site - in contrast to Germany - flooded sewers are to be expected as a rule, since the sewers cannot be shoved clear at short notice for condition recording due to administrative reasons. Therefore it was decided that instead of a sledge, which mainly works in half-filled, a raft construction should be used.

Also on the software side, adaptations for the next practical test have been worked out. The data generated from the first practical test was viewed and the export was adjusted accordingly. Graphic elements were inserted into the report of the sewer section and further data were added. Furthermore, the master data of the main sewer objects and the condition standard of the software were extended by condition codes in accordance with the specifications of the Chinese project partners.

The software itself and its operating concept was very positively received by the Chinese project partners, so that no fundamental reconstruction works were necessary.

7.4.4 Second Demonstration of the Inspection Procedure

For the second practical test in April/ May 2018, the findings and suggestions from the first demonstration were implemented. With regard to the on-site demonstration,

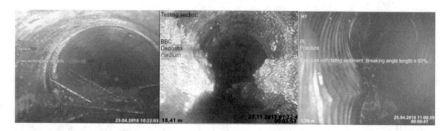

Fig. 7.26 Results of inspection (bluemetric)

the localisation (German, English and Mandarin) was implemented as far as possible and the inspection functions were programmed with the provided hardware.

The Chinese standard provided by our project partner FiW could be implemented for the condition documentation. This standard replaced the simplified condition recording, which was shown during the first demonstration. Since no formats for data transfer of the sewers to be inspected could be provided, the software was extended in such a way that the generation of sewer sections with corresponding master data is possible directly on site. A project management allows the temporal grouping of the work. As no exchange interface could be named during the analysis of the existing systems in China (in Germany a defined XML file is used for this purpose, DWA M150 or ISYBAU), it was decided to display the data via CSV file. In addition, a report with master data, sewer section graphics and the generated images per condition is available for each sewer section as a PDF file. Video files and image files have also been provided. In addition, a license-free "viewing version" was integrated for viewing the data. This enables the precise time-controlled navigation of the documented conditions in the video. Photos and reports on the inspected object can also be opened. No additional software is necessary to view the results of the sewer section inspection. The viewer can be started directly from a CD or USB drive. The resulting export files were successfully presented and have attracted great interest. The sewer sections inspected during the on-site demonstration showed enormous damage. The documented obstacles and deformations could be directly identified as the cause for the very high water level (Fig. 7.26).

In addition to the sewer sections in a residential area, a sewer section in an industrial area was also investigated. Here the raft construction created by JT-elektronik was used.

After on-site demonstration, the project was presented 2018 at the IEexpo in Shanghai. For this purpose, an imagine-film was commissioned, which shows the problems of wastewater disposal and introduces the involved companies.

7.4.5   Second Assessment of Procedure

As already experienced during the first practical test, the examined sewer sections showed very high water levels and deposits, so that an optical inspection was hardly feasible. However, the documentation produced was able to give those responsible a direct illustration of the problem. The documentation software was used successfully and, despite adverse conditions, recorded the condition of the sewers very clearly.

The presented technology can thus be applied to document the condition and causes of acute and flow-preventing objects by means of conditions, photos, video and position that can be electronically evaluated by machine. This can be used to plan measures. However, a comprehensive documentation of the condition is not possible due to the water level and the lack of cleaning. Rather, an initial documentation can be carried out so that critical problems can be localised and appropriate immediate measures can be taken to ensure that the wastewater is drained off. The crucial task of sewer cleaning in order to maintain the desired condition and remove contamination and deposits must be preceded by an inspection. The inspection for assessment and evaluation can only be carried out comprehensively in this way.

7.5 Recommendations for Actions

(See Tables 7.3, 7.4 and 7.5).
Operational planning

Table 7.3 Evaluation of the deposit situation

Task of Operation	Goals/ Recommendations for Action	Tools
Data recording	• Assignment of deposition data according to shaft and sewer section • Digital data recording	• TV-inspection • Sewer data base / KIS • Standardised protocols • Mobile recording devices
Data editing	• Evaluation according to drainage systems and pollution classes • Visualising the data (Photo/ Video)	• Sewer database/ KIS • GIS-systems
Assessment	• Set-up of a Demand plan for cleaning	• Assessment by • Demand plan for cleaning • Visualisation/ GIS

Table 7.4 Assessment of the situation

Operational Task	Goals/ Recommendations for Action	Tools
Data recording	• Manhole and section of sewer-related assignment of the damage situation • Digital data recording	• TV-inspection • Sewer database/ KIS • Standardized protocols • Mobile recording devices
Data editing	• Evaluation according to specified standards for condition • Visualisation of the date (photo/ video)	• Sewer data-base/ KIS • Creation of a data exchange interface • Training of specialists • Engineer assessment
Assessment	• Establishment of an inspection plan • Priority determination	• Engineer assessment • Inspection plan • Visualisation/ GIS

Table 7.5 Operational planning

Operational Task	Goals/ Recommendations for Action	Tools
Organisation and relevance	• Organisation of a fast availability of capacities to eliminate incidents • Planning of the daily/ weekly cleaning and inspection operations	• Flushing and operation plan
Service and costs	• Generation of performance reports on cleaned manholes and sewers • Recording of operating costs • Identification of the need for rehabilitation and a rehabilitation strategy	• Sewer data-base / KIS • Daily reports • Mobile data recording • Statistical surveys

References

1. Deutsches Institut für Normung (2017) DIN – kurz erklärt. https://www.din.de/de/ueber-nor men-und-standards/basiswissen, zuletzt abgerufen am 15 July 2017
2. DIN 1986–3 (2004) Entwässerungsanlagen für Gebäude und Grundstücke – Teil 3: Regeln für Betrieb und Wartung. Beuth Verlag GmbH, Berlin
3. DIN EN 13508–2 (2011) Untersuchung und Beurteilung von Entwässerungssystemen außerhalb von Gebäuden – Teil 2: Kodiersystem für die optische Inspektion. Beuth Verlag GmbH, Berlin
4. DIN EN 752–5 (1997) Entwässerungssysteme außerhalb von Gebäuden – Teil 5: Sanierung; Deutsche Fassung EN 752–5:1997. Beuth Verlag GmbH, Berlin
5. DWA-M 149–5 (2010) Zustandserfassung und -beurteilung von Entwässerungssystemen außerhalb von Gebäuden – Teil 5: Optische Inspektion. Hennef: DWA.
6. DWA-M 150 (2010) Datenaustauschformat für die Zustandserfassung von Entwässerungssystemen (April 2010). DWA, Hennef
7. Lecher K, Lühr H-P, Zanke UCE (eds) (2001) Taschenbuch der Wasserwirtschaft. Springer Vieweg, Wiesbaden
8. (LfU) Bayerisches Landesamt für Umwelt (Hrsg.) (2016) Leitfaden zur Inspektion und Sanierung kommunaler Abwasserkanäle. LfU, Augsburg
9. Meteoblue (2016) Klima Modell Jiaxing. https://www.meteoblue.com/de/wetter/historycl imate/climatemodelled/jiaxing_china_1805953, zuletzt abgerufen am 15 July 2017
10. STATISTA (2013) Länge des Kanalnetzes in Deutschland. http://de.statista.com/statistik/daten/ studie/152743/umfrage/laenge-deskanalnetzes-in-deutschland-im-jahr-2007/
11. Guo S, Shao Y, Zhang T, Zhu DZ, Zhang Y (2013) Physical modeling on sand erosion around defective sewer pipes under the influence of groundwater. J. Hydra. Eng. 139(12):1247–1257
12. Zhou Y, Zhang Y, Tang P, Chen Y, Zhu DZ (2013) Experimental study of the performance of a siphon sediment cleansing set in a CSO chamber. Water Sci Technol 68(1): 184–191.
13. Guo S, Zhang T, Zhang Y, Zhu DZ (2013) An approximate solution for two-dimensional groundwater infiltration in sewer systems. Water Sci Technol 67(2):347–352
14. Zhou Y, Zhang Y, Tang P (2013) Field performance of self-siphon sediment cleansing set for sediment removal in deep CSO chamber. Water Sci Technol 67(2):278–283